THE EXPERIMENTER'S COMPANION
A GUIDE AND REFERENCE TO
THE ASPECTS OF RESEARCH AND EXPERIMENTATION

THE EXPERIMENTER'S COMPANION
A GUIDE AND REFERENCE TO
THE ASPECTS OF RESEARCH AND EXPERIMENTATION

Richard B. Clements, Solution Specialists

ASQC Quality Press
Milwaukee, Wisconsin

THE EXPERIMENTER'S COMPANION
A GUIDE AND REFERENCE TO THE ASPECTS OF RESEARCH AND EXPERIMENTATION

Richard B. Clements

Library of Congress Cataloging-in-Publication Data

Clements, Richard B.,
    The experimenter's companion/Richard B. Clements.
        p.   cm.
    Includes bibliographical references (p. 128).
    ISBN 0-87389-114-7
    1. Experimental design. I. Title.
    T57.37.C58   1991
    001.4'34—dc20   91-10438
                                                                     CIP

10987654321

ISBN 0-87389-114-7

Acquisitions Editor: Jeanine L. Lau
Production Editor: Mary Beth Nilles
Set in 12-point Times by Patricia L. Coogan.
Cover design by Laura Bober.
Printed and bound by BookCrafters.

For a free copy of the ASQC Quality Press Publications Catalog,
including ASQC membership information, call 800-952-6587.

Printed in the United States of America

ASQC Quality Press
611 East Wisconsin Avenue
Milwaukee, Wisconsin 53202

This book is dedicated to every student who has had the courage to study a new subject, and to every manager, technician, engineer, worker, and executive who has said, "I don't know, but I'm going to find out."

Table Of Contents

Preface

Since 1980, the need to conduct industrial experiments has increased rapidly as companies seek ways to create higher quality products at the lowest possible costs. Experimentation is the ultimate tool for achieving this goal.

This book was written for anyone interested in conducting experiments in the industrial environment. It shows you how to create an experimental design, collect data, analyze results, make improvements, and report results.

In addition, this book is written in a clear, straightforward style to make experimental work as simple and logical as possible. We have no tolerance for jargon or mathematical complexity. We are not performing proofs, we are educating a new audience of experimenters.

As you read this book, keep in mind that this is a guide. If any of the subjects excites your interest, please turn to the Recommended Readings section and seek out additional books on the topic.

We wish you well in learning one of the most powerful job skills you can obtain.

Before You Begin

You will need some materials to complement this book. The most important is a hand-held calculator with a square root key. Even better would be a calculator that can perform statistical summaries. This will assist you in the analysis of experiments.

It is possible to use statistical software while reading this book, but we recommend that you follow the formulas by hand for better comprehension of the subject matter.

The coverage of specific analysis is deliberately kept light. You are encouraged to consult other books on the topics discussed herein. This book is intended as a sampler of the techniques used in a good experiment. After reading this book, you should be able to conduct a few simple experiments. More importantly, you will be ready for serious study of the subject.

Chapter One
The Experiment

In this section we talk about:
- what an experiment is
- why we experiment in industry
- how an experiment is conducted

What is an experiment?

When I was a young boy, I would perform an experiment every winter. I would go to the nearby pond and check to see if the ice on it could support my weight.

By direct observation I could not tell if the ice was thick enough to hold my weight. Therefore, I performed a test. I would pick up a rock about the size of a loaf of bread and toss it onto the ice. If it made a cracking noise on impact, the ice was too thin. If it landed with a delightful "thump," the ice could hold my weight.

This experiment was possible because observations and past tests had taught me about the fundamental nature of the ice.

Thus, an experiment is our use of tests to explore what is not directly observed. This is as true in industry as it is on an ice pond. A plastic injection molding process may create parts that shrink too much. We cannot directly observe what is occurring to cause the shrinkage, so we apply a wide variety of methods to ferret out the cause.

This process is also called an investigation. In the investigation of a problem we may observe the machine, examine the SPC charts, keep new statistical records, and perform experiments. An experiment is just one part of a larger investigation. It is critical that we do not focus on just the experiment.

Thus, an experiment is a series of tests performed to determine why a particular situation occurs. In our example above, we wonder why the plastic parts produced by a molding machine shrink. Experience and reference works may tell us that several factors may be responsible, such as the mold temperature, injection speed, type of plastic resin being used, and so on.

An experiment will determine which of these factors affect shrinkage the most. In addition, it will give us enough information to create mathematical models of how variations in these factors affect the rate of shrinkage. This is critical. With such models we can predict the results of specific settings for the factors, or better still, we can optimize the molding process to fit our needs.

Why experiment?

Experiments are costly, time-consuming, sometimes successful, and almost always frustrating to conduct. So why would we be so interested in conducting an experiment?

We experiment to enhance our competitive advantages through efficiency. Let me explain. Efficiency is the use of a resource at the lowest possible expense. The resource, in this case, is knowledge. With superior knowledge, a company can easily out-compete its rivals.

An experiment increases knowledge through testing. For example, let's assume that I own a welding machine. The strength of the welds it creates varies. Through experimentation I can determine which factors affect this variation the most. For the purpose of illustration, we will assume that I am testing the following factors:

- welding wire used
- voltage
- duration of voltage
- welding speed

If my experiment finds that the welding speed contributes 78 percent of the variation in the quality of the welds, I have gained powerful knowledge. With this information I can make informed changes in the parameters of the welding process. I can also change product designs to take advantage of the welding machine characteristics. In addition, I can predict the performance of the welding process if new materials are introduced.

As we shall see, this information allows a company to be very competitive. One way is through increasing the efficiency of a design. An engineer can tighten tolerances that are critical to the success of the product. Then the cost of the final product is reduced while its quality has been increased.

Another path toward competitiveness through experimentation is the search for the causes of product or process problems. By finding the "whys" of a situation, a company can eliminate the causes of problems from a system or product through design. Later, when we introduce the works of Dr. Genichi Taguchi, we will see that experiments can also increase the robustness of a product or process.

4

How to experiment

What follows are the steps taken in any experiment. These go beyond conducting the experiment. They focus on making our findings useful and, hopefully, profitable.

1. Investigate the situation.
2. Form a central question.
3. Design the experiment.
4. Gather data.
5. Perform analysis.
6. Build a mathematical model.
7. Verify results.
8. Take action.

These eight steps represent a great deal of time and effort. Therefore, we recommend that you exhaust all possible "easy answers" before conducting an experiment. These simple answers are found by using obvious steps, such as referring to a machine's instruction booklet or applying a law of physics.

Step 1: investigate the situation

A critical part of any experiment is to conduct an exhaustive literature review and investigation of the situation. Perhaps you have been assigned to improve the quality of automotive rearview mirror assemblies. Your first job would be to pin down definitions of "quality," "improvement," and "assemblies."

Assuming you find the number of visual defects in the rearview mirror assemblies is too high, your next job would be to review the literature in your field to see if an answer to your problem already exists. If not, then you will want to interview the people involved in the assembly process to see if obvious answers are being overlooked.

Step 2: form a central question

The results of your intensive investigation of the situation should result in a well-formed question that leads to your experimental design. For example, you may have found that several different assembly procedures are used for the rearview mirrors. Your central question may be: "Which assembly methods lead to the fewest number of visual defects in the rearview mirrors?"

Step 3: design the experiment

The design of an experiment involves selecting which tests to perform and deciding the order in which to conduct them. The idea is to test factors you feel are important and whether any of those factors, alone or in combination, significantly change the variation of the response variable.

One of the hardest achievements is to ensure that your central question dictates your experimental design. Many times experimenters latch onto a favorite method and fit all problems into its experimental design. This must be avoided at all costs.

If we are interested in finding associations, we may choose to do regressional studies. On the other hand, we may find that several factors have to be tested for their contributions to the number of defects found. In such a case, we may need a factorial experiment.

In the next chapter, we will examine the issues and methods of experimental design in greater detail.

Step 4: gather data

Once you have established a design, you implement it and collect the resulting data. In the example of the rearview mirrors, you would count the number of visual defects for each new experimental assembly method.

Step 5: perform analysis

Next, apply statistical tests to the data to look for significant interactions or associations. In most of the experiments in this book, Analysis of Variance (ANOVA) is used. Many industrial experiments are analyzed using ANOVA. However, don't overlook the use of other tests for your own experiments when the situation warrants different analysis.

Chapter 3 will discuss ANOVA in detail.

Step 6: build a mathematical model

The initial analysis will tell you the effect of each factor on the variation in results. However, for true competitiveness, you should build a mathematical model of the dynamics of the system under study. This allows you to make predictions about the outcome of future actions and to optimize designs.

Chapter 5 shows some of the mathematical models that can be formed using data from an experiment.

Step 7: verify results

The results of any experiment usually are not believed until a confirming test is performed. In many experimental situations in industry, you can confirm an experiment with a production run. Another alternative is to run a second, confirming experiment to see if the results match those predicted by the mathematical model.

Step 8: take action

This is perhaps the most overlooked of the steps. No one in industry should conduct an experiment unless its results are going to be put to work. The suggested new settings for a machine or new tolerances developed by an experiment must be applied in the manufacturing area. The idea is to track the improvement and document the success of the experiment in the real world. Without this dedication to tracking performance and using experimental results, your company will be wasting a great deal of money.

Your role as an experimenter

As the person actually designing and conducting an experiment, you have several responsibilities. The most important of these is to establish an experimentation policy for your company that will assure the quality and integrity of each experiment.

This policy must establish a group of managers to approve proposed experimental work and to review the results so that actions can be implemented. In addition, this policy must assure that complete and detailed records are kept.

Other responsibilities include:

1. Ensuring that the research you wish to do has not already been done.

2. Assuring that an experiment is the proper response to the problem at hand.

3. Checking to see if an experiment will yield practical results.

4. Talking to at least ten people involved in the process before experimenting.

5. Controlling variables as strictly as possible.

6. Maintaining a written record of your actions and results.

7. Publishing your results, either as a report to the managers or as an article in a technical journal.

8. Seeking review by other authorities in the field. It is especially important that your design and the results of analysis be double-checked for accuracy and validity.

9. Seeking continuous education. As an experimenter you must seek out new methods and ideas in the field. This book only scratches the surface of experimental methods.

Where to begin

People working on their first experiment are often confused on where to begin.

The first step is to select a situation to investigate. This is usually the first source of confusion. Within any company there is always a host of problems occurring with new products, old products, and ongoing production processes. From the list of potential situations to investigate, you must apply a standard set of priorities to select the best targets for investigation.

Once you have selected a particular situation, you should examine it in detail. For example, you boss may instruct you to investigate the failure of chair legs at your office furniture plant. Your first job would be to pin down definitions and numbers for what is meant by "failures." Perhaps you find that customers return five percent of the office chairs the company makes because the chairs have broken or cracked legs. Clearly, this problem demands an investigation.

Common sense is a powerful tool to use in all aspects of investigating and experimenting. Once a situation has been well defined, common sense should dictate the next step. The chair leg example provides us with some possible steps to be considered.

The first possibility is to examine a sample of the returned chairs to determine the location and types of breaks. Simple data-gathering techniques, such as defect location maps or c-bar charts, might suggest a common cause for the breaks. For example, you might find that all of the breaks occur on the two back legs. Such a strong clue should not be overlooked. It could indicate a simple and direct answer to the situation and prevent the need to experiment.

Let's look at another possible course of action. Suppose that you found the breaks occurred fairly equally between legs and were all halfway between the ground and the point where the legs connected to the chair. By testing the strength of the material in the legs and then using simple physics, you might determine that the legs are incapable of supporting a leveraged load.

Therefore, the entire chair requires redesign. You could use quality function deployment or another form of product development to improve the entire chair. Experiments would then assure that the new specifications and tolerances would meet customer needs.

Yet another possible course of action could take you to a personal computer. Using a database program, you might enter the data reported for each chair returned. By sorting and counting factors such as the customer's location and use of the chair versus reason for return, you may find that the situation has other undetected problems. For example, you may find that poor fabric appearance was a common second reason for returning the chair. Thus, you would want to broaden your investigation.

The lesson to be learned here is that you should begin with a strong and clear definition of the situation under investigation. Your own common sense should determine the next logical step to take. In most cases, this will involve a series of tests, examinations, and interviews. After a while, you may find that an experiment is necessary because there is no obvious solution to the problem.

When an experiment is called for, you should take some time to consider which factors to test. After you select the factors, the design and analysis of the experiment can proceed.

Selecting factors

One of the most important steps in designing an experiment is selecting the factors to study. Basically, you have two groups of factors to name. The first are the experimental factors, those that you will be manipulating. These are also called the independent variables. The second are the response factors you will measure. These are called the dependent variables.

If you remember the Pareto principle, there should be hundreds of factors to choose from, but only a vital few that really make significant changes to the response. Therefore, you should spend a good deal of time investigating the factors to test.

Some books will tell you to "brainstorm" with the people involved in the process to come up with a list of factors. We agree, but we also recommend that you perform a literature search and talk to experts to complete the list.

SPC, multi-var charts, engineering studies, and other statistical tools can also go a long way in identifying important factors. For example, the reaction plan log from a SPC chart may have frequent notes about the need to adjust the RPMs of an automatic screw machine to correct for variations in final diameters. This provides you with a vital clue in your search.

When selecting factors to manipulate, choose those that are known to have a real effect and that can be easily controlled by the production department. For example, if you are testing a plastic injection machine, factors such as injection speed, mold temperature, and percent regrind are known to have real effects on the process. A factor such as "vendor" will not work well because you cannot pick and choose vendors easily.

Once you have chosen the experimental factors, you must check their history statistically. Using cross-tabulation and auto-regression software, compare each factor against the other factors. If two experimental factors are found to correlate, then we have to assume that they are not independent from one another. Corrective action or a redesign of the experiment must take place.

Select response factors for their practicality. They should be hard, variable-scale factors where possible. Thus, factors such as volts, shrinkage, size, and defect counts make good response factors. The percent yield is an example of a poor response variable. It does not tell us the quality of the rejected material.

Please note that you can test a single product on several response scales. There is no rule against checking plastic parts for shrinkage, sinks, color, cracks, and visual appeal at the same time. You merely run an analysis for each response factor.

Attribute data, such a types of defects, can be coded for experimental analysis. One approach is to rate the severity of a defect across a scale. This converts pure counts into variable data.

Summary

Although an experiment is not the only tool available for an investigation, it is one of the most powerful available. Its ability to probe the unseen and to find the "whys" of a situation can be extremely valuable to a company seeking a more competitive position. Not only will experiments help solve current problems, but they can also improve the design of products.

However, the use of experimentation demands the application of a set of well-defined rules. An experimenter has to be precise, demanding, meticulous, and well-educated. Above all, an experimenter must use common sense in every step of the experiment.

Chapter Two
Experimental Design

In this section we talk about:
- the issue of validity and reliability in an experiment
- the threats to validity
- classic experimental designs
- quasi-experimental designs
- full-factorial designs
- fractional factorial designs

Validity and reliability

An experiment is a structured investigation. In industry, hundreds of variables can affect a process. In an experiment we control as many of those variables as possible, holding them constant. We change a few critical variables of interest by design and study their effect on the process.

An experimental design is the "structure" in a structured investigation. It spells out the exact sequence of operations to take place during an experiment. It also has to achieve two very important goals. The first is to make the experiment valid, so that others will trust your study. The second is to make the experiment reliable. A reliable experiment can be repeated, and this helps others have confidence in your results.

You can achieve reliability in your experiment by running confirming experiments or production runs. We will discuss these in chapter 5. In this chapter, the major emphasis will be on creating an experimental design that will be robust in withstanding the "threats to validity." These are occurrences that can bias your results or create a false interpretation.

The threats to validity

The twelve threats to validity are:

1. **History**–Events occur between measurements to add to the variation observed.

2. **Maturation**–Parts, tools, and materials wear down or change over time.

3. **Testing**–Your first set of measurements affects the results of later measurements.

4. **Instrumentation**–Calibrations of measuring instruments change or are not properly set.

5. **Regression**–Usually occurs when samples have been selected because of their unusual nature.

6. **Bias**–The selection of samples, tests, measurements, and so on because of an experimenter's specific interest.

7. **Experimental mortality**–The loss of parts or samples.

8. **Interactions**–The effect of an interaction may be confounded with the effect of a factor.

9. **Reactive effect**–Initial testing may artificially reduce variation during later measurements.

10. **Reaction to bias**–An interaction between a bias and the experimental variables.

11. **Experimental arrangement**–An experiment in which the variables being tested would not reflect real-life situations.

12. **Multi-treatments**–The effect of one treatment variable remains even after the treatment is no longer being applied.

For further information on these threats, consult Stanley and Campbell's *Experimental and Quasi-Experimental Designs for Research* (Rand McNally, 1963).

A single example illustrates how these threats can devastate your experiment. Suppose that you tested a heat treatment oven at temperatures ranging from 500° to 550° F and found that treatment at 500° increased the strength of a part. The problem is that your oven can operate only as high as 400° to be cost-effective. Your testing range of temperatures does not match the reality of the situation. Therefore, the validity of your results can be called into question.

Another example would be an experimenter who did not randomize the order of experimental treatments. Let's suppose that this experimenter was testing the effect of three vendors' sheet steel against the final dimensional stability of stamped parts. If the press operator stamped 30 parts from vendor 1 steel, then 30 from vendor 2, and finally 30 from vendor 3, the validity of the experiment would be questionable. How could we distinguish changes in dimensional qualities from the changes in the machine over time?

The thirteenth threat

After years of conducting experimental studies in industry, we have learned of a thirteenth threat to validity: internal politics. This occurs when management decides that your experimental results did not provide the answers they wanted to hear.

For example, a recent study found that structural problems in a product were the result of low-cost materials used in its construction. However, fixing the problem would require the company to return to the costlier material it previously purchased. The problem is that one of the managers had suggested the switch as a cost-reduction measure. Needless to say, he doesn't want to hear that he made the wrong decision.

Experience will teach you that this type of situation is often repeated within a company. A good experimental design will defeat the first 12 threats to validity. However, the 13th threat is only countered with a strong policy on experimentation supported by a top management staff which is not afraid of learning.

Experimental design is a counter-measure to the threats

Experimental design is a process. Its objective is to create a set of instructions on how an experiment will be conducted. This includes which factors will be varied, how they will be varied, and when they will be varied.

The goal of experimental design is to form an experiment that will obtain valid and reliable information that can be used to make objective conclusions about a particular situation. As we have seen, there are a dozen different threats standing in the way of this goal. Luckily, several approaches have been developed to defeat most of the threats.

Balanced against the need for valid and reliable results is the need for an efficient experiment. It is possible to require millions of tests to learn every aspect of a new product's performance. However, the years of research involved would keep the product off the market so long that it would no longer be competitive. Therefore, the experimenter in industry usually will be faced with choosing a design that satisfies both the need for information and the time-sensitive world of business.

In addition to choosing an experimental design, you must also take steps to assure the usability of your data. One step is to experiment only with factors that can be easily controlled. Humidity in a production area can only be controlled through the use of expensive machinery. Therefore, it makes little sense to study its effect if you cannot do anything to change it.

A second step is to randomize the order of an experiment and the assignment of subjects whenever possible. By randomizing, you avoid several threats to validity, such as bias, history, and maturation.

Another step is to test for interactions and correlations between factors. When these exist they can "confound" your results—that is, give a false reading of significance.

Experimental designs

There are thousands of established experimental designs. Usually, an experimenter only needs to choose the best one for a particular situation.

Before we look at the families of designs, here are a few definitions.

- **Factor**–The experimental variables.
- **Response**–The measured outcome of an experiment.
- **Level**–The settings of a factor.
- **Treatment**–A combination of experimental factors at specific levels.
- **Run**–A single test of a combination of factors.

In an experimental design, the experimenter lists the factors to be studied. Each factor is tried at selected levels. Each combination of factor levels is called a treatment. The number of runs in an experiment is determined by the number of factors and the number of levels for these factors.

For example, a chemist may want to test the effect of heat on the speed of a chemical reaction. For the factor of "heat," he or she might select three levels, such as 200°, 250°, and 300° F. We would call this a design with one factor at three levels. To test each level, at least three runs are necessary.

The speed of the reaction at each of the three temperature levels would be the response data. However, the response is not just the result of the changing of temperatures. Factors outside of the experiment, such as the purity of the chemicals used, will contribute to the variation of the response. A good experiment is one in which the experimenter accounts for and controls the effects of outside factors.

The effect of other variables outside of the experimental design is called experimental error. The power of an experiment is that this error can be estimated by making replications and repetitions of the experiment. Thus, an experimenter can determine if a variable was missed that should have been included as a factor in the experiment. Otherwise, the sensitivity of the experiment can be lost to this outside "noise."

When an experimenter finds an assignable cause for experimental error, he can change the design of the experiment to increase the validity of the results. For example, an experiment on three different molding machines implies that three different operators will be involved. Since the experimenter can control the assignment of operators, he will want to randomly assign an operator to a machine for each run.

When a source of experimental error is beyond control, blocking is a good countermeasure. To "block" an experiment requires including the nuisance factor in the design. Then the remaining sets of experimental factors are represented within each "block."

As an example, suppose that a cookie bakery is testing batches of fig newtons. The experiment will take two days, and the temperature of the cooling area will change dramatically from morning to afternoon. The increased heat will require the cookies to cool longer, which will affect their final taste and texture. Therefore, the blocking factor will be room temperature. The experimenter will perform each test in both cool and warm conditions. This will allow the experimenter to determine the effect of the changes in cooling room temperatures.

19

To fully understand how an experimental design is created and used, it is best to examine the different "families" of design.

The "classic" experimental design family

The most basic experimental design is called the "classic" design. It has also been called the one-factor-at-a-time design. In this design, we randomly divide our subjects into two equal groups. One group is the experimental group, and the other serves as a control group. An example can illustrate how this design works.

First, let's take a look at the actual design written in experimenter's shorthand (see table 2.1).

Table 2.1: The classic experimental design.

$$R \quad O_1 \quad X \quad O_2$$

$$R \quad O_3 \quad \quad O_4$$

Time---->

The top row represents the experimental group. The "X" indicates that this group will receive the experimental treatment. The control group, represented by the second row, will not receive the treatment.

For example, let's suppose that a woman purchased a box of fertilizer sticks for her houseplants. She wants to know if the sticks really make a difference. The factor being studied is the effect of the sticks. To conduct the experiment she would first randomly divide her plants into two groups. This is what the "Rs" in the design refer to. Then she would make an observation of each group (this is what the "Os" indicate.) In this case, she would measure the height of all the plants.

Then she would insert the fertilizer sticks into the soil of one group of plants. This would be the experimental group. The control group would not get the sticks. However, both groups would still receive equal amounts of water and sunlight. After a few weeks, she would measure plant heights again and test to see if the experimental group had grown faster than the control group. These are the last two "Os" in the design.

Not only would the analysis tell her whether the fertilizer had a significant effect, but the data could also be used to create a mathematical model of the effect. Therefore, she could predict plant growth on a regular diet of the sticks.

Other "classic" designs

For testing a single factor and making simple comparisons, the "classic" method is unbeatable. Much of the early work in scientific research used this approach. The early work on genetics is an excellent example. In addition, there are dozens of designs that use the one-factor-at-a-time approach. These are listed in table 2.2 for your examination. Each has its particular application and function.

Table 2.2: More examples of classic designs.

The Solomon 4-Group

R	O_1	X	O_2	
R	O_3		O_4	
R		X	O_5	
R			O_6	

Posttest-Only Design

R		X	O_1
R			O_2

Hovland's Delayed Effect Test

R	O_1	X	O_2	
R	O_3		O_4	
R	O_5	X		O_6
R	O_7			O_8

The Solomon 4-Group design uses four groups instead of the two groups of the classic design. The two groups receiving the treatment are still the experimental group, and the other two groups are the control. The design does not make an initial observation of the last two groups. This avoids the effect of testing on the subjects.

For example, you may want to study the effect of better lighting in the assembly area on the number of assembly defects. In the "classic" design, you would have to first observe the workers before installing new lights for half of the area. Unfortunately, it is possible that the workers will change their normal behavior knowing they are being observed.

The Solomon 4-Group design corrects for this possibility by not making any observations within two groups until the treatment has been applied. In our example, not only would we check the number of defects between the lit and unlit groups, but we would also check the number of defects between the observed and unobserved groups. This second analysis would indicate whether a testing effect has occurred.

The Posttest-Only design is similar but is used in situations where the first set of observations cannot be made or a known testing effect might occur. The most likely application of such a design occurs when a totally new technique or method is introduced. We cannot pretest because the subject matter is new.

Hovland's Delayed Effect Test is suited to those situations where a change may take time to be noticed, such as the introduction of a new piece of equipment. The product being made is still the same, but it may take several weeks for the machine's crew to "get up to speed."

Quasi-experimental designs

In addition to these "classic" designs, there are some quasi-experimental derivatives (see table 2.3). These violate one or two threats to validity to cope with unusual experimental circumstances, such as the inability to draw a random sample or the time-dependent nature of the data collected.

The Non-Equivalent Group design is used when it is not possible to randomly assign subjects into two groups. In some industrial experiments, contractual agreements or material demands do not permit random assignments. The experimenter must then take pains to observe any differences in the test groups before proceeding with the experiment.

Table 2.3: Quasi-experimental designs.

Non-Equivalent Group Design

$$O_1 \qquad X \qquad O_2$$
$$O_3 \qquad\qquad\quad O_4$$

Simple Time-Series

$$O_1 \quad O_2 \quad O_3 \quad O_4 \quad X \quad O_5 \quad O_6 \quad O_7 \quad O_8$$

Multi-Group Time Series

$$O_1 \quad O_2 \quad O_3 \quad O_4 \quad X \quad O_5 \quad O_6 \quad O_7 \quad O_8$$
$$O_9 \quad O_{10} \quad O_{11} \quad O_{12} \qquad O_{13} \quad O_{14} \quad O_{15} \quad O_{16}$$

Equivalent Time Sample

$$O_1 \quad O_2 \quad X \quad O_3 \quad O_4 \quad X \quad O_5 \quad X \quad O_6 \quad O_7$$

The time-series experiments watch a process over time and then introduce a treatment. After that, the experimenter makes many more observations over time. SPC is a form of this design. At regular intervals an operator takes a sample of parts and makes an observation. For example, he might count the number of defects. A new production method would then serve as the treatment. The SPC system continues to make observations. Any real changes in the number of defects will drive the plotted points out of statistical control.

Again, each of these designs has its place and purpose. Most of the time, the "classic" designs are not used in the industrial environment because of the difficulties in controlling all but one variable in a production process.

The "modern" designs

The revolution in experimental methods came during the first part of the 20th century. Mathematicians developed methods to test more than one factor at a time. These new designs came to be called "full-factorial" experiments. With them, you can test many factors at many levels simultaneously.

A factorial experiment tests two or more factors at two or more levels each. The simplest illustration would be two factors at two levels. Let's assume that you are testing a new drilling machine. Ideally, this machine will drill 0.125-inch holes into quarter-inch steel. You have to choose between two types of drill bits, carbide-tip or diamond-tip. The drill bits represent factor A. The two types of bits are the two levels for factor A.

Factor B is the speed of the drill. The manual that came with the machine recommends 1200 rpm for quarter-inch steel. Your engineer recommends 1500 rpm. Therefore, the two speeds will represent the two levels of factor B.

The actual design creates all combinations of drill bits and speeds. In the shorthand of experimentation, this is called a 2^2 experiment. This represents the number of levels for each factor, raised to the power of the number of factors. The resulting design on paper might look like table 2.4, a simple box.

Table 2.4: A full-factorial design as combinations of all factor settings.

	Factor A	
Factor B	**Carbide**	**Diamond**
1200		
1500		

As you can see, this forms a design of four "squares" or experimental combinations. Thus, a 2^2 design creates four experimental runs, or combinations.

To conduct this experiment, you would randomly choose the order in which each combination would be run. For example, one possibility might be to run the squares in a 1-4-2-3 order. The data collected during each run are written into the corresponding square. To analyze the effect of factor A, we would look at the averages for the columns. For factor B, we would analyze the row averages (see table 2.5).

Table 2.5: Data filled into the full-factorial design.

Factor A

Factor B	Carbide	Diamond	Row Average
1200	0.124 1	0.118 2	0.121
1500	0.126 3	0.127 4	0.1265
Column Average	0.125	0.1225	

An additional piece of information we can obtain from a factorial design is the interaction between the two factors under study. Perhaps a certain combination of speed and drill bit produces the best results. In chapter 3 we will examine the formal methods of analysis. However, it is important to point out that all analysis should begin with a casual look at the actual numbers.

In our example, we are seeking a hole size of 0.125 inches. The carbide-tip drill seems to create a stable hole size near this target at both 1200 and 1500 rpm. Thus, we begin to suspect that the carbide-tip drill is the one we desire to use in production.

Factorial experiments do have a problem, however. As you increase the number of factors and levels, the designs become large and ungainly. For example, a simple plastic molding machine might have twelve control knobs, each of which can be set to one of five positions. A full-factorial design for this machine would be 5^{12}. This represents almost a quarter of a billion combinations.

25

Full-factorial experiments are usually excellent and economical when dealing with no more than five or six factors at a time. Larger scale factorial experiments usually limit themselves to two or three levels for each factor.

Also, by limiting to two levels for each factor, it becomes much easier to write out the design of the experiment by creating a matrix of settings. Three factors tested at two levels each (2^3) can serve as an excellent example (see table 2.6).

Table 2.6: Three factors at two levels each.

	Factors		
Run	A	B	C
1	–	–	–
2	+	–	–
3	–	+	–
4	+	+	–
5	–	–	+
6	+	–	+
7	–	+	+
8	+	+	+

Notice how low and high settings for a factor correspond to "+" and "–"? This gives us a bonus. To calculate the effect of, say, factor A, we add the results that correspond to the pluses and minuses under the A column. To calculate an interaction between factors A and B, we can create a new column of "AB." Using the product of the A column times the B column, we can quickly calculate the column for the A×B interaction. We can then analyze this column as if it were also a factor.

This method of analysis is explored fully in the next chapter.

Randomized block design

We can use the full-factorial design described above to counteract the effect of a nuisance factor outside the experiment. For example, suppose that you were testing the effect of injection speed on the shrinkage of molded plastic parts. Although you are using a single machine and mold for the test, you produce the part in many colors. You create each of the colors by altering the chemical composition of each plastic resin. Therefore, the differences in the resins represent an outside factor which you suspect may change the rate of shrinkage for the parts.

To both counter and explore this possibility, you incorporate the colors as another factor in the design. However, you do not randomly assign all of the combinations. Instead, each "block" formed by a different color contains a complete set of the other factor combinations randomized for the experiment (see table 2.7). This will ensure that we detect both the effect of color and the injection speed, the factor in which we were originally interested.

Table 2.7: Injection speed vs. plastic color experiment.

Injection Speed

Plastic Color	1.5 sec.	2.0 sec.	2.5 sec.
Blue	2	3	1
Black	1	2	3
White	1	3	2
Red	2	1	3

Order each block is run

1 = First   2 = Second   3 = Third

The result is a single factor randomized block design. By blocking the experiment according to the color of the plastic, we can test for the effect of this outside factor. The factor is considered to be outside of the experiment, because in practice you would not be able to select a single color for the product. Customer requirements have demanded the use of multiple colors.

Other factorial designs

As the number and availability of factors change, so do the requirements on a design. Therefore, there are dozens of possible designs available. For example, if we did not have enough material to try every combination of factors inside a block, we would use an incomplete block design.

There are also cases in which the resetting of a factor is very expensive or time-consuming. Take the case of a molding machine that normally runs at 360° F. If molding temperature was one factor in the experiment, we might expect to change it back and forth between two levels, such as 360° and 400°. Unfortunately, each change may require well over an hour for the machine to stabilize at its new setting. Therefore, we waste a lot of time in the experiment.

The solution is to use a design called split-plot. In this design, the molding temperature would be set to 360° and all of the other combinations of factors would be run. Then the heat would be increased to 400° and the combination of factors run again.

There are designs such as the Latin Square that allow you to fold one factor into a two-way design. Designs can be combined to create composite layouts. Additional factors can be "nested" within a design. The point is that the conditions of your situation should dictate which design to use. You should never select one design and try to change the situation to fit the design.

Fractional factorial designs and Taguchi

By the 1940s it was becoming clear that the factorial design did have its limitations. As the number of factors under study increased, the number of possible interactions increased exponentially. The problem was that most of these interactions were between three or more factors. The likelihood of these interactions was very remote. Thus, the efficiency of the experiment suffered.

One of the solutions proposed for such situations was to use a fractional design. This is literally what it sounds like: only a fraction of the full-factorial combinations is used. However, creating a fractional design is difficult. The resulting design must be mathematically balanced and statistically representative.

The technique of creating a fractional design depends upon the assumptions you are making about possible interactions. Additional factors are added to a full-factorial design by confounding them with interactions. By confounding, we mean selecting an interaction column that we believe has little effect. A new factor is assigned to that column. Thus, the effects of the new factor and the previous interaction are combined, or confounded. The assumption is that the effect of, say, a three-level interaction will be small because such an interaction is unlikely. Therefore, we assume the resulting effect is mostly the result of the assigned factor.

Although this seems to be an awkward and dangerous way to design an experiment, in reality the method is fairly effective. The result is an experiment that can test many factors with very few runs (see table 2.8).

Table 2.8: A fractional-factorial design.

Factors

Run	A	B	C
1	–	–	+
2	+	–	–
3	–	+	–
4	+	+	+

Thus, by the end of the 1940s you had to be a sophisticated statistician to use the fractional designs. Remember, computers were not available at that time. The alternative was to conduct the larger but simpler experiment to analyze a full-factorial design.

This was the state of affairs when Dr. Genichi Taguchi came onto the scene. As it turns out, the need for a compromise and the events of history would give birth to the third approach to experimentation, the Taguchi approach. We will examine the Taguchi approach in more detail in chapter 4. For now, it is important to realize that Dr. Taguchi's philosophy represents a real change in experimentation. The methods he used are based on those discussed above.

Taguchi virtually created a cookbook of fractional designs and a method of quickly assigning factors and two-way interactions to the proper columns. The result was a new and controversial method of experimental design.

Repetition and replication

Repetition refers to the situation in which each experimental run takes several samples. For example, we set the machine to a particular experimental combination and run off several parts. The amount of error in these samples is said to be "within experiment error," or what Taguchi calls secondary error (e_2).

By setting the size of the repetition samples to be taken, an experimenter has direct control over the amount of sampling error within an experiment. For example, a larger repetition sample will decrease the sampling error estimate and increase the ability to distinguish real differences between factors.

Replication refers to repeating the experimental combinations and taking samples with each change. In other words, we would try each experimental combination in a design and take a sample. Then we would replicate (re-run) the experimental combinations and take additional samples. The error that this creates represents the accuracy of the settings we used and the effect of outside factors. It is called "between experiment error." Taguchi calls this primary error (e_1).

By increasing the number of replications, an experimenter can obtain a better estimate of the experimental error. This helps assure the sensitivity of the experiment.

In the end, most experimenters will find that they have to use repetitions and replications along with randomization to assure a sound design for their experiment. Unfortunately, the economics of the situation usually restrict the experimenter from assigning high numbers of repetitions and replications.

In most of the examples of experimental analysis that follow, single repetitions are used. This is to ease the burden of calculation of analysis and should not be seen as a model of a proper experiment. In practice, you will need to use and analyze both repetitions and replications (see chapter 4).

Chapter Three
The Analysis of Experiments

In this section we talk about:
- the Analysis of Variance (ANOVA)
  and the test of a hypothesis

Review of the statistics involved

There are two statistical methods to review before talking about experimentation. The first is the correct method of drawing a sample. The second is the test of hypothesis.

Sampling

The key word in sampling is "random." A random sample means that all members of a population have an equal chance of being selected to be in the sample.

To draw a random sample, the best method is to have a computer generate a series of random digits. Usually, these will be fractions between the numbers zero and one; for example:

0.1395
0.3954
0.9822
0.5899

To draw a random sample, you multiply each of these numbers by the number of subjects in the population. For example, suppose that I have a batch of 300 cylinders in a box and I want to randomly select 10 cylinders. I would generate 10 random digits and multiply each by 300. The answer, once rounded off, would tell me which part to select.

$$0.1395 \times 300 = 41.85$$

Rounded off, this would be 42. Therefore, I would select the 42nd cylinder in the box. I would repeat this procedure until I had selected all ten parts.

Hypothesis testing

John Cleese of the comedy troupe Monty Python once made an interesting observation: "In an experiment, we don't test to see if our ideas are right; we test to see if they are wrong."

This is precisely what a test of a hypothesis is all about. We begin by forming our ideas into a specific hypothesis about a situation—for example: "I believe that machine #1 creates parts with less variation than machine #2." Then we test this idea statistically.

The test of a hypothesis is the keystone of an experiment. The procedures outlined below make the test valid and reliable. This will make your results more believable to others.

In simple terms, making a test of your hypothesis is like betting with a stranger. You need to lay down some ground rules and state your bet clearly to assure a fair playing field. In the test of a hypothesis, you begin by stating a "null hypothesis."

A null hypothesis is a statement of your idea as if nothing unusual was occurring—for example: "I believe that machine #1 and machine #2 both produce parts of equal variation." The tests you conduct will not "prove" this statement to be true; instead they will see if it might be false. In other words, if the test fails to find a difference, then you accept the null hypothesis.

The actual process is:

1. State the null hypothesis.

2. Select a level of significance for the statistical tests.

3. Pick a statistical test.

4. Identify the sampling distribution.

5. Form a decision rule.

6. Collect and analyze data.

7. Accept or reject the null hypothesis.

An example will illustrate this method in greater detail.

Step 1: state the null hypothesis

Let's suppose that we do indeed have two automatic screw machines. Each has been assigned the job of creating a cylinder casing that is one inch in outside diameter. We are interested in whether the two machines create the same diameter part on average.

An example of a null hypothesis might be, "There is no difference in the average diameter of parts from machine #1 and machine #2."

Step 2: select the level of significance

The level of significance refers to the chance that the statistical test will reject a hypothesis that in reality is true. This is also called the Type I Error, or Alpha Error. In industrial experiments, this level is usually set between 10 percent and 1 percent Alpha Error. In our example, we will use a 5 percent level of Alpha Error.

However, the Alpha Error is not the only type of error we must consider when testing a hypothesis. There is also the chance that we would accept a false null hypothesis. This is called Type II Error, or Beta Error.

Taken together, these two types of error demonstrate the need to "design" an experiment. With a fixed sample size, as you increase the level of significance for Alpha, the risk of a Beta Error increases. The experimenter needs to balance the two risks by selecting the sample size that will produce acceptable levels of risk in the test. The larger the sample size, the less will be the risk that the sample estimates will be misleading.

Step 3: pick a statistical test

The exact statistical test to use in hypothesis testing is dictated by your particular situation. In our example, we are comparing two sample means, and we are using small sample sizes of five parts each. This dictates that we select the following t-test for comparing two means with an equal number of parts selected in each sample.

$$t = \frac{\bar{x}_1 - \bar{x}_2}{\sqrt{\dfrac{s_1{}^2}{n_1} + \dfrac{s_2{}^2}{n_2}}}$$

where, $\bar{x}_1$ = the first average

$\bar{x}_2$ = the second average

s_1 = the first standard deviation

s_2 = the second standard deviation

n_1 = the first sample's size

n_2 = the second sample's size

Most elementary statistics books and technical references, such as Juran and Gryna's *Quality Planning and Analysis,* outline which test to use for a particular situation.

Step 4: identify the sampling distribution

The sampling distribution is the probability distribution used under the assumption that the null hypothesis is true. In this case, it is the "t" distribution. The "t" distribution represents the probability curve for small samples. Appendix A contains the "t" distribution information. This probability information is critical for the next step.

Step 5: form a decision rule

A closer examination of Appendix A shows us that we can look up a particular value for a 0.05 Alpha Error for a sample of five parts. Because we don't care whether one machine in particular is better than another, we use what is called a "two-tail test." Furthermore, we calculate the degrees of freedom for this test. We will discuss this concept in detail later, but for right now the formula is,

$$df = n_1 + n_2 - 2$$

or,

$$df = 5 + 5 - 2 = 8$$

So, when we look up the value at 8 degrees of freedom, we find the number 2.306. This number is used to form the decision rule. Any result of the t-test that lands between +2.306 and −2.306 is considered acceptable. In other words, an answer within this range can occur purely by chance.

Step 6: collect and analyze data

For the sake of illustration, we will assume that five parts were randomly selected from each machine (see table 3.1).

Table 3.1: Five samples from two machines.

	Machine 1	Machine 2
	1.002	1.007
	1.005	1.006
	1.000	1.009
	1.002	1.007
Average	1.0022	1.0068
Std. Dev.	0.0018	0.0015

Upon first examination of the data, we can see that the sample from machine #2 has about the same variation as the sample from machine #1 (see figure 3.1). However, the average for each machine is clearly different. We can see this difference in the figure. Nevertheless, to demonstrate how a test of a hypothesis would detect a significant difference where it wouldn't be so obvious, we proceed to test this data statistically by applying the t-test.

Figure 3.1: An illustration of the zones of rejection.

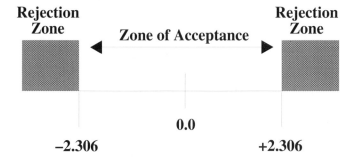

37

$$t = \frac{1.0022 - 1.0068}{\sqrt{\frac{0.0018^2}{5} + \frac{0.0015^2}{5}}}$$

or,

$$t = -4.39$$

Step 7: accept or reject the null hypothesis

The resulting t-value of -4.39 easily falls outside the acceptance range of our decision rule. Therefore, we reject the null hypothesis that these two machines produce parts that have equal average diameters.

Analysis of Variance (ANOVA)

Analysis of Variance (ANOVA) is one of the most commonly used methods of data analysis in industrial experiments. Its popularity is due to its ability to sort out the effect of several factors on a single response. In other words, if you wanted to test six factors on a stamping press against a particular dimension, the ANOVA can quickly tell you which factor creates the most variation in part size.

An example can illustrate the power of an ANOVA. Let's suppose that you have three gluing machines that attach fabric to wood. The quality of the bond is measured as the pounds of pull necessary to strip off a one-inch-wide strip of fabric. Thus, "pounds" is your response factor.

In this example, we are testing at a 95 percent level of significance. Our null hypothesis is that the three gluing machines produce equal average fabric bonds.

$$H_o: \mu_1 = \mu_2 = \mu_3$$

Because this is a destructive test, you want to take small samples and compare the average pull strength of each machine to the others. An ANOVA can perform this test. To see how, let's look at the data you collect. The machines have the names Alpha, Beta, and Gamma (see table 3.2).

Table 3.2: Strip force results from three machines.

Sample	Alpha	Beta	Gamma
1	11	13	8
2	12	14	11
3	13	12	7
4	9	12	15
5	11	13	9
Totals	56	64	50
		Grand Total	170

As you look at the raw data, your first reaction may be to calculate averages for each of the machines and to compare these. A better alternative would be to plot the data to see if an obvious difference exists (see figure 3.2).

Figure 3.2: Plotting the data from three machines.

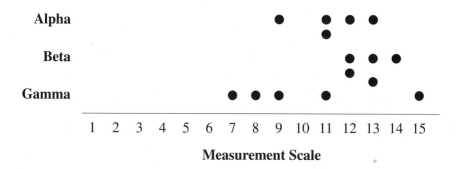

This simple exercise can reveal interesting comparisons. For example, the averages for the machines are close to one another. However, the variation in Gamma's production seems to be wider than the variation in the other two machines. If we were to put all of the data together, we would obtain a fairly normal distribution of the production from the three machines working together.

Now, observe that the data vary for all three machines. This is called the total experimental variation. Inside this total variation are two components. The first is the "experimental noise." It represents variations occurring because of sampling, missing factors, and replications. The second component is the "signal" generated by the difference between factor levels.

For example, each machine represents a level of the factor "machine." The ANOVA will test whether the signal produced by changing these levels is stronger than the background "noise" of the experiment. If it is, then we know that the factor has a significant effect. That is, we will know that all three machines do not produce identical effects.

The analysis

Because only one factor is involved, this is called a one-way ANOVA. It is the simplest of the ANOVAs, yet it is an excellent illustration of how an ANOVA works.

The effects of the experiment are "decomposed" using a method called the sum of squares. Eventually, we will translate these sums of squares into variances to perform the actual analysis. In the meantime, you will find the hardest part of the ANOVA being the calculation of the sum of squares (called "SS" in statistical shorthand).

For this example, we will break down the data into its three parts: total variation, variation from the factor, and error variation (the "noise" between samples).

First, we calculate a correction factor. This is used in the remaining sum of squares calculations.

$$CF = \frac{(\sum x)^2}{N}$$

where $\sum x$ = the sum of all the data

N = the number of numbers for the entire experiment

or,

$$CF = \frac{170^2}{15} = 1926.7$$

Then we calculate the sum of squares for the total experiment

$$SS_{total} = \sum x^2 - CF$$

where $\sum x^2$ = the sum of each number squared.

Using the above data,

$$SS_{total} = (11^2 + 13^2 + \ldots + 9^2) - 1926.7 = 71.3$$

Next, we calculate the effect of the factor and the error.

$$SSA = \sum \frac{col.\ totals^2}{n} - CF$$

so,

$$SSA = \frac{56^2}{5} + \frac{64^2}{5} + \frac{50^2}{5} - 1926.7 = 19.7$$

and,

$$SSE = \sum x^2 - \sum \frac{col.\ totals^2}{n}$$

thus,

$$SSE = (11^2 + 13^2 + \ldots + 9^2) - \left[\frac{56^2}{5} + \frac{64^2}{5} + \frac{50^2}{5}\right] = 51.6$$

To conduct the actual ANOVA, we summarize these results in what is called an ANOVA Table (See table 3.3).

Table 3.3: An ANOVA Table being formed.

Source	SS (sum of squares)
A	19.7
error	51.6
Total	71.3

This is only the first step in completing the table. Next we add a column for the degrees of freedom. "Degrees of freedom" represents the amount of information the data contain. In most cases, this will be one less than the number of levels for a factor, or one less than the sample size.

For example, the degrees of freedom for the whole experiment is one less than the total number of data points, or $15 - 1 = 14$. The degrees of freedom for any factor is one less than the number of levels of the factor. Our factor had three levels (Alpha, Beta, and Gamma), thus $3 - 1 = 2$ degrees of freedom. The degrees of freedom for error is the difference between what the factors use and the total for the experiment. In this case, $14 - 2 = 12$ degrees of freedom.

We add another column to the ANOVA Table for the variance. The variance is the sum of squares divided by the degrees of freedom. For example, the SSA is 19.7 with two degrees of freedom. Thus, the variance is $19.7 \div 2 = 9.85$. This is repeated for each row, except for the total sum of squares.

$$V_A = \frac{SSA}{df_A}$$

The resulting ANOVA Table appears in table 3.4.

Table 3.4: Variance is added to the ANOVA Table.

Source	SS	df	V
A	19.7	2	9.85
error	51.6	12	4.30
Total	71.3	14	

The final step in this analysis is to calculate the F-ratio between each factor and the error. This is the ratio between signal and noise. In our example, the variance for factor A is divided by the variance for error.

$$F_A = \frac{V_A}{V_{error}}$$

or,

$$F = \frac{9.85}{4.3} = 2.29$$

To find if this is significant, we would refer to the F-ratio table for 95 percent confidence, located in Appendix B. The numerator has two degrees of freedom and the denominator has 12 degrees of freedom. The critical value is 3.89. Our answer of 2.29 is below this, so the machines are producing bonds of essentially equal strength.

ANOVA for multiple factors

For any full-factorial experiment or a Taguchi design, you will have to use an ANOVA for multiple factors. Luckily, a design convention used for both types of experiments makes this calculation relatively easy.

To begin with, both types align their factors into columns. For an example, we will look at the case of a manufacturing company having problems with one of its metal lathes. This machine is supposed to turn a high-technology alloy rod into a drive shaft for a vehicle produced under a defense contract. The problem is that the cutting tools are breaking too frequently.

The company's engineer decides to test three factors at two levels each.

Factor A = turning speed (High = 400 rpm, Low = 300 rpm)

Factor B = tool used (Fred's or Dwayne's)

Factor C = cutting oil mixture (40 percent or 20 percent)

The resulting (2^3) design appears in table 3.5, which shows the number of pieces made before the tool broke.

Table 3.5: A full-factorial design.

Run	Factors			Results
	A	**B**	**C**	
1	–	–	–	87
2	+	–	–	62
3	–	+	–	124
4	+	+	–	58
5	–	–	+	89
6	+	–	+	34
7	–	+	+	63
8	+	+	+	37
			Total	554

The calculation of the sum of squares becomes much easier through use of this table, because the results by level are paired with the plus and minus signs in each factor column. Let's complete the ANOVA for this data and see how this works.

First, we calculate the correction factor as before.

$$CF = \frac{(\Sigma x)^2}{N} = \frac{(554)^2}{8} = 38{,}364.5$$

Then we calculate the sum of squares for the total experiment.

$$SS_{total} = \Sigma x^2 - CF = (87^2 + 62^2 + \ldots + 37^2) - 38{,}364.5 = 6{,}203.5$$

Now, to calculate the effect of each factor, we pair the plus and minus signs with the corresponding results. Let's begin with factor A. (See table 3.6.)

Table 3.6: ANOVA Table for factor A.

Run	Factor A	Results
1	–	87
2	+	62
3	–	124
4	+	58
5	–	89
6	+	34
7	–	63
8	+	37

Thus,

$$A_{low} = 87 + 124 + 89 + 63 = 363$$

and

$$A_{high} = 62 + 58 + 34 + 37 = 191$$

The formula for the sum of squares becomes

$$SSA = \frac{A_{low}^2}{n} + \frac{A_{high}^2}{n} - CF$$

thus,

$$SSA = \frac{363^2}{4} + \frac{191^2}{4} - 38{,}364.5 = 3{,}698$$

We can repeat this same procedure for the other columns in the design.

$$SSB = \frac{B_{low}^2}{n} + \frac{B_{high}^2}{n} - CF$$

Thus,

$$SSB = 12.5$$

and,

$$SSC = \frac{C_{low}^2}{n} + \frac{C_{high}^2}{n} - CF$$

Thus,

$$SSC = 1{,}458$$

Again, the degrees of freedom for each factor is one less than the number of levels ($df = k - 1$, or $df = 2 - 1 = 1$). The degrees of freedom for the entire experiment is one less than the number of data points ($df = n - 1$, or $df = 8 - 1 = 7$).

Interactions

At this point we would have four degrees of freedom left in our analysis. These four represent the four possible types of interactions between three factors.

A × B	B × C
A × C	A × B × C

As discussed before, interactions occur when the combination of two or more factors creates a significant effect, though each individual factor has very little effect. Ammonia and bleach are great cleaners by themselves; however, when mixed they give off a deadly gas and could explode.

In industrial experiments, interactions are usually of the greatest interest to the design engineer. An interaction could reveal a combination that can maximize a situation for greater profitability. In our example, a combination of two or more of the factors we are testing might dramatically reduce the number of tools breaking.

To discover this, we need to expand our analysis to test for interactions. In the 2^3 design, we have three two-way interactions and one three-way interaction to test. We can perform this analysis in the same way we calculated the sums of squares for the three factors; all we need are columns in our design for interactions.

We create these columns by multiplying the signs in the existing three columns. For example, we can take the factor A column and multiply it by the factor B column to create the A×B interaction column (see figure 3.3).

Figure 3.3: Creating an A×B interaction column.

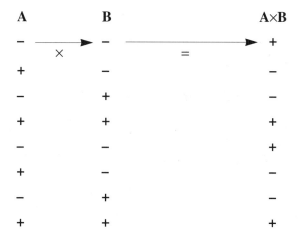

We repeat this procedure for the other possible combinations. The result is seven columns in our design for analysis (see figure 3.4).

Figure 3.4: The full-factorial design for three factors.

Run	A	B	C	A×B	A×C	B×C	A×B×C
1	−	−	−	+	+	+	−
2	+	−	−	−	−	+	+
3	−	+	−	−	+	−	+
4	+	+	−	+	−	−	−
5	−	−	+	+	−	−	+
6	+	−	+	−	+	−	−
7	−	+	+	−	−	+	−
8	+	+	+	+	+	+	+

The calculation for the sums of squares proceeds as before. We match the plus and minus signs in each column with the results obtained from the experiment. To make this procedure easier, we can create what is called a level sum table (see table 3.7).

Table 3.7: The level sum table.

	A	B	C	A×B	A×C	B×C	A×B×C
High (+)	191	282	223	271	282	249	312
Low (−)	363	272	331	283	272	305	242

Notice how the level sum table helps us organize the calculation of the sums of squares. For example, to calculate the sum of squares for the $A \times B$ interaction,

$$SS_{A \times B} = \frac{A \times B_{low}^2}{n} + \frac{A \times B_{high}^2}{n} - CF$$

In other words,

$$SS_{A \times B} = \frac{283^2}{4} + \frac{271^2}{4} - 38{,}364.5$$

$$SS_{A \times B} = 18$$

Likewise,

$$SS_{A \times C} = \frac{A \times C_{low}^2}{n} + \frac{A \times C_{high}^2}{n} - CF$$

$$SS_{A \times C} = 12.5$$

$$SS_{B \times C} = \frac{B \times C_{low}^2}{n} + \frac{B \times C_{high}^2}{n} - CF$$

$$SS_{B \times C} = 392$$

$$S_{A \times B \times C} = \frac{A \times B \times C_{low}^2}{n} + \frac{A \times B \times C_{high}^2}{n} - CF$$

$$SS_{A \times B \times C} = 612.5$$

We add these sums of squares into the ANOVA Table. To complete the table, we need to calculate the sum of squares for the error. The easy way to do this is to realize that the sum of squares for the error will be what remains after we have calculated the factors.

In other words,

$$SSe = SS_{total} - SS_A - SS_B - SS_C - SS_{A \times B} - SS_{A \times C} - SS_{B \times C} - SS_{A \times B \times C}$$

or,

$$SSe = 6{,}203.5 - 3698 - 12.5 - 1458 - 18 - 12.5 - 392 - 612.5 = 0$$

Note how we end up with no estimate for the experimental error. This occurred because we performed no replications of the experimental runs. Thus, our example is also a demonstration of the importance of performing replications.

The degrees of freedom for the sum of squares for error matches the above formula:

$$df_{error} = df_{total} - df_A - df_B - df_C - df_{A \times B} - df_{A \times C} - df_{B \times C} - df_{A \times B \times C}$$

or,

$$df_{error} = 7 - 1 - 1 - 1 - 1 - 1 - 1 = 0$$

In this example, without replications, the degrees of freedom for the error is zero. In our next example you will see how the addition of replications will increase the degrees of freedom in our estimate of the experimental error.

With the sums of squares now calculated, we can complete the ANOVA Table for this experiment (see table 3.8).

Table 3.8: The ANOVA Table for interactions.

Source	SS	df	V
A	3,698.0	1	3,698.0
B	12.5	1	12.5
C	1,458.0	1	1,458.0
A×B	18.0	1	18.0
A×C	12.5	1	12.5
B×C	392.0	1	392.0
A×B×C	612.5	1	612.5
error	0.0	0	
Total	6,203.5	7	

Pooling

Now a new problem has arisen. In our previous calculation for ANOVA we divided the error variance into the variance for each factor to create F-ratios. Unfortunately, we have no error variance in table 3.8. The solution to this problem is to pool out insignificant effects.

In our example we can quickly see that the sum of squares for factor B and interactions A×B and A×C are very small and probably insignificant. We will assume that their small effects are really the result of the "background noise" or error of the experiment. Therefore, we can pool their sums of squares together with the three degrees of freedom they possess.

This pooling goes on the ANOVA Table into a new source we call the eror of the pooled variance. Table 3.9 illustrates how this is done.

Table 3.9: Pooling with the ANOVA Table for interactions.

Source	SS	df	V	F
A	3,698.0	1	3,698.0	258.00
B	12.5	1	12.5	
C	1,458.0	1	1,458.0	101.72
A×B	18.0	1	18.0	
A×C	12.5	1	12.5	
B×C	392.0	1	392.0	27.35
A×B×C	612.5	1	612.5	42.73
error	0.0	0		
Total	6,203.5	7		
pooled error	43.0	3	14.33	

By dividing the sum of squares for the pooled error (43) by the three degrees of freedom, we create our needed error variance. In this case, the error variance is 14.33, and it can be divided into the surviving variances to create the F-ratios.

The F-ratios were created with one degree of freedom in the numerator and three degrees of freedom in the denominator. Using Appendix B, we find that the critical F value is 10.13. Factors A and C, as well as interactions B×C and A×B×C, are significant.

What these results tell us is that the speed of the lathe and the cutting oil mixture (factors A and C) contribute significantly to the tool life. Another important piece of information is that the type of tool used makes very little difference. This is good news because it will not restrict the number of tool sources for the company.

Examining the interaction effects, we find that the B×C and A×B×C interactions seem to be significant at the 95 percent level as well. This would lead the experimenter to make further investigations of these interactions. Chapter 5 discusses some of the ways this could be done.

Putting ANOVA to work

From this point we can use the analysis for further studies and to make economically wise changes to the process. The first step would be to find the best settings for this process. This is accomplished using effect diagrams and mathematical models.

An effect diagram shows the change in the response factor as an experimental factor is changed. For example, let's look at the effect of turning speed on the life of the tool. We would calculate the average tool life for the High setting (400 rpm). This would be the numbers associated with the plus signs in the "A" column.

$$\textit{High average for A} \;=\; \frac{(62 + 58 + 34 + 37)}{4} \;=\; 47.75$$

The Low average would be the number associated with the minus signs in the "A" column.

$$\textit{Low average for A} \;=\; \frac{(87 + 124 + 89 + 63)}{4} \;=\; 90.75$$

Clearly, the lower speed of 300 rpm is the preferred setting (see figure 3.5). However, confirmation runs would have to bear this out.

Figure 3.5: Effect diagram for factor A.

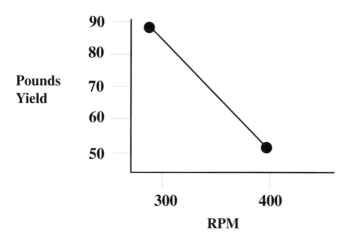

The alternative would be to form a mathematical model of the effect of each significant factor and optimize the formula for the best results. We will talk about this more in chapter 5.

If the results prove interesting and the experimenter wishes to optimize the settings, the choice is to run another experiment on the significant factors using three or more levels. This will reveal whether the effects are linear or quadratic. Chapter 4 will discuss the importance of non-linear effects in engineering a better product or process.

Analysis for replications

As we saw in the previous example, without replications we have to pool to get an error variance. Another disadvantage is that the sensitivity of an experiment is usually too low when only one replication is performed. For example, as a rule of thumb, it's a good idea to have at least six degrees of freedom in your error sum of squares.

Let's suppose that an adhesive manufacturer is testing new mixing procedures for a particular type of glue. The critical result is to create a product with a very stable viscosity. To test the viscosity, each batch is sampled by dipping 10 inches of a metal rod in the mixture. The viscosity is reported as the number of seconds it takes for the glue to run off the rod. The company wants a viscosity of 30.

The engineer in charge of this project has selected three factors to test.

- the percentage of solvent used (5 percent or 5.5 percent)
- the temperature of the mixing tank (110° or 120° F)
- the percentage of solids in the mixture (2 percent or 3 percent)

Obviously, a 2^3 factorial design will work just fine for these conditions. The engineer will mix the eight possible combinations of the factors. Then he will replicate the experiment by mixing another set of eight possible combinations.After testing the sixteen resulting batches, he obtains the information shown in table 3.10.

Table 3.10: The full-factorial design for replications.

Run	Factors			Results		Total
	A	B	C			
1	−	−	−	30	31	61
2	+	−	−	29	30	59
3	−	+	−	25	25	50
4	+	+	−	26	27	53
5	−	−	+	31	31	62
6	+	−	+	32	30	62
7	−	+	+	24	25	49
8	+	+	+	26	24	50
				Grand Total		446

A level sum table is generated using the procedure outlined in the previous example (See table 3.11).

Table 3.11: The level sum table for replications.

	A	B	C	A×B	A×C	B×C	A×B×C
Low (−)	222	244	223	220	223	227	225
High (+)	224	202	223	226	223	219	221

To calculate the sum of squares for the factors, interactions, and the total experiment, we use the same equations as before. However, note how the n in each equation is increased to reflect the increased information from replication.

$$SSA = \frac{A_{low}^2}{n} + \frac{A_{high}^2}{n} - CF$$

thus,

$$SSA = \frac{222^2}{8} + \frac{224^2}{8} - 12{,}432.25 = 0.25$$

We can repeat this procedure for the other columns in the design.

$$SSB = \frac{B_{low}^2}{n} + \frac{B_{high}^2}{n} - CF$$

Thus,

$$SSB = 110.25$$

and,

$$SSC = \frac{C_{low}^2}{n} + \frac{C_{high}^2}{n} - CF$$

Thus,

$$SSC = 0$$

$$SS_A \times B = 2.25$$

$$SS_A \times C = 0$$

$$SS_B \times C = 4.0$$

$$SS_A \times B \times C = 1.0$$

The only real change in our method of calculation is how the sum of squares for the error is calculated. To make this calculation we have to create two different sums of squares for the total effect of the experiment. Our previous total will now be called the SST_2. The squares of cell totals will become SST_1. The difference between these two totals will be the sum of squares for the replication error.

$$SST_1 = \frac{1}{r} (Cell_1^2 + Cell_2^2 + \ldots + Cell_n^2) - CF$$

thus,

$$SST_1 = \frac{1}{2} (61^2 + 59^2 + \ldots + 50^2) - 12{,}432.25 = 117.75$$

and,

$$SST_2 = \Sigma x^2 - CF = (30^2 + 31^2 + 29^2 + \ldots 24^2) - 12{,}432.25 = 123.75$$

The sum of squares for the repetition error is calculated as

$$SS_{error_2} = SST_2 - SST_1$$

or,

$$SS_{error_2} = 123.75 - 117.75 = 6$$

For the degrees of freedom,

$$df_{error\,2} = df_{SST_1} - df_A - df_B - df_C - df_{A \times B} - df_{A \times C} - df_{B \times C} - df_{A \times B \times C}$$

Since the total experiment has 15 degrees of freedom, the equation becomes:

$$df = 15 - 1 - 1 - 1 - 1 - 1 - 1 - 1 = 8$$

All of this information can be placed into an ANOVA Table for the complete analysis. Again, we create a column for variance and a column for the F-ratios. Since each F-ratio is created with one degree of freedom in the numerator and eight in the denominator, our critical F-value is 5.32.

Table 3.12: The ANOVA Table for the replication experiment.

Source	SS	df	V	F
A	0.25	1	0.25	0.33
B	110.25	1	110.25	147.00
C	0.00	1	0.00	0.00
A×B	2.25	1	2.25	3.00
A×C	0.00	1	0.00	0.00
B×C	4.00	1	4.00	5.33
A×B×C	1.00	1	1.00	1.33
error$_2$	6.00	8	0.75	
Total	123.75	15		

The results in table 3.12 clearly show that factor B, the temperature of the mixing tank, is the critical factor in controlling viscosity. A slight interaction between temperature and the percent solids used is also evident. The next step would be further investigation of at least factor B at more than two levels to create the optimal viscosity.

In the next chapter we will discuss the Taguchi approach. Remember that, when testing factors at three or more levels using the Taguchi designs, you may analyze your information in the same fashion.

Chapter Four
The Taguchi Approach

In this section we talk about:
- the historic background of Dr. Taguchi and his methods
- the use of orthogonal arrays
- the use of linear graphs
- designing experiments for factors at two levels
- designing experiments for factors at three levels
- modifying the designs for various levels of factors

Background

Dr. Genichi Taguchi is a product of history. At the close of World War II, U.S. General Douglas MacArthur took control of Japan. His General Headquarters there set the stage for the progress Japan was to achieve over the next 40 years.

One of the major projects the General Headquarters assigned was the improvement of the Japanese telephone system. The headquarters staff wanted Japan to copy the system developed by AT&T's Bell Laboratories in the United States. Unfortunately, the Nippon Telephone and Telegraph research and development division was only 1/50th the size of AT&T's unit. Furthermore, the staff estimated that it would take 20 years to bring the Japanese telephone system up to American standards, yet they had only five years to finish the job. Dr. Taguchi, a staff member, put forward a proposal to speed the research process. He recommended using the fractional designs and standardizing the research method for all personnel. Four years later, the telephone development job was done, and Dr. Taguchi was famous in Japan.

For the next 20 years, Dr. Taguchi held classes on his new method in many Third World countries. Eventually, in 1980, he gave his first class at Bell Labs. In 1984, the American Supplier Institute founded permanent courses in his methods in the U.S.

Taguchi's approach to experimentation is controversial in the U.S., and it has attracted both praise and criticism worldwide. However, in Japan, tens of thousands of experiments conducted in industry each year use this technique. In the U.S., both the automotive industry and the Department of Defense have embraced his methods.

Taguchi's philosophy

Taguchi's approach is based on philosophical as well as scientific foundations. The best example of this is his definition of quality:

*Quality is the loss imparted on society after a product
is shipped.*

In other words, companies have a greater responsibility than merely generating profits and providing jobs. They also have to create the highest quality goods at the lowest price possible to maximize the benefits to the customer.

Accordingly, Taguchi has published extensively on various topics related to quality and productivity. We will limit our discussion to his view on experimentation.

Similar to his Western counterparts, Taguchi sees experimentation as the study of variation. However, his view of variation is different. He sees variations in a product as causing a loss to society. He divides the loss into two segments.

1. The loss caused by functional variations; in other words, a product failing to function as specified will cause problems for the producer and the consumer.

2. Loss due to harmful effects of the product, such as an electrical circuit that shorts out and starts a fire.

Taguchi sees the engineer as the designer of quality. The goal of engineering, in Taguchi's eyes, is to dampen the negative effects of a product and to build the product as close to optimal specifications as possible. An example can illustrate this idea.

Let's suppose that an engineer has specified that a power supply must produce a 100-volt current. If it supplied 100 volts at all times, its quality would be perfect. Unfortunately, it is impossible to create a supply that has perfect quality. Other factors prevent this from happening. Taguchi calls these interfering factors "noise" and divides them into three types.

1. **Outer noise** covers variations caused by environmental factors such as heat, humidity, input voltage, atmospheric pressure, and so on.

2. **Inner noise** refers to variations caused by the decay of the product materials. The way that resistors in a circuit increase in resistance over time is an example of inner noise.

3. **Between-product noise** refers to the natural variations that occur between manufactured products. The Cpk index is an example of how we currently measure this type of noise.

Thus the role of the engineer is to create countermeasures to these noise factors by improving the design of a product or process. This approach is different than the one used in the West. Western engineers tend to search for the cause of a problem so that they may eliminate the cause. Taguchi tends to search for ways to dampen the effect of the problem. In other words, he seeks to make the product robust against the sources of variation.

Such a product would be designed so that the variations that occur during manufacturing, such as changes in material quality or the tools being used, have little effect on the final quality of the product.

Taguchi divides this effort into three areas, collectively called "off-line quality control."

1. **System design**–similar to manufacturing engineering. Experiments determine the selection of materials, method of production, plant layouts, and machines of greatest utility.

2. **Parameter design**–experiments find the optimal settings on a machine or process to achieve the highest quality at the lowest cost.

3. **Tolerance design**–This involves finding the best tolerance settings for a design.

In summary, Taguchi focuses on economy, market, and mathematics. His method is technological, not theoretical. The production of a high-quality product begins with more emphasis on optimizing before manufacturing begins. His techniques are designed for practical, real-world situations.

What makes Taguchi different?

A lot of the techniques used by Taguchi practitioners are based on conventional experimental design techniques, such as the fractional factorial design and the ANOVA Table. Many people ask, then, what is so different about the Taguchi approach that makes it so controversial.

The first of these differences is the definition of quality. Dr. Taguchi sees it as the loss to society after a product is shipped. Westerners see it as conformance to specifications during production. The Taguchi definition shifts the manufacturer's responsibility to include the world outside of the plant, as well as examining all quality questions in the light of economics.

The philosophy of experimental design is also a sharp contrast. In the West, the emphasis is on difficult mathematical models addressing a few factors at a time, studying interactions, and setting critical levels of significance to find basic cause-and-effect relationships as an addition to general scientific knowledge.

In comparison, Taguchi uses only some of these methods. In addition, he has designed a simple-to-use system (at the expense of statistical purity) that can test dozens of factors at one time using forms of the fractional factorial design. Unlike traditional fractional designs, this system seeks to eliminate or combine interactions into a single factor. Taguchi sees interactions as "too difficult" to control directly, while Western experimenters tend to emphasize the study of interactions as the "really interesting" part of the experiment.

The use of linear graphs is another point of controversy. Taguchi uses linear graphs as an aid to assigning factors to an orthogonal array. The practitioners of fractional factorial design spend much more time carefully assigning factors to assure that two important effects are not confounded in the results.

When Taguchi uses the F-test during ANOVA, the idea of "critical levels" is rejected in favor of seeing a factor's response as a continuum. Thus, an experiment is designed to provide engineering information. Taguchi sees the F-ratio as a continuous response and treats it as such.

The selection of factors, the actions taken after an experiment, and the setting of tolerances in a design are all based on a decidedly Eastern view of economic considerations. Taguchi sees it as far better to make a good design with inexpensive parts than to make an excellent product with costly, complex components. The loss to society is less with the first product.

In short, Taguchi's approach is an inductive process. Designs and processes are optimized before manufacturing begins. Taguchi has even pointed out that a company with a strong research and development department does not need to use his methods. His method of experimenting is intended for the company that cannot obtain the quality of materials to match its competitors.

The actual method

Let's assume that a researcher at a chemical processing plant is testing batch mixing methods. The result he is looking for is the pounds of fertilizer that the process yields. He is studying three factors that can be controlled during the batch mixing: the temperature of the vat, the time the chemicals are allowed to rest before mixing, and the amount of agitation used during mixing. These are summarized as,

Factors:
A = temperature
B = time
C = agitation

For each of these factors, the researcher has selected two possible settings. Each represents the high or low of the "normal" operating range of the process.

A_{high} = 140°
A_{low} = 100°
B_{high} = 20 minutes
B_{low} = 10 minutes
C_{high} = high agitation
C_{low} = low agitation

The normal analysis of this experiment would, perhaps, be a conventional full-factorial experiment (2^3), and we would have eight experimental combinations to test. However, to illustrate a Taguchi experiment, these same factors will be tested with only four experimental runs (see figure 4.1).

Figure 4.1: Comparison of Taguchi array with full-factorial.

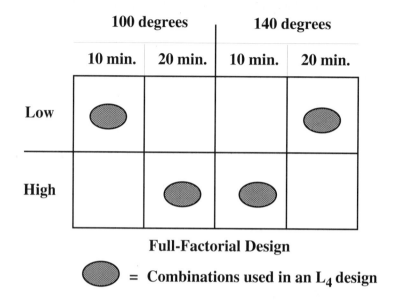

Full-Factorial Design

= **Combinations used in an L₄ design**

CAUTION: In this particular example, there will be only one replication of the experiment. This is done to keep the illustration and its analysis simple for greater clarity of how a Taguchi experiment works. In actual experiments, you would have several replications to increase the accuracy of the error estimate and thus increase the sensitivity of the experiment. We will point out the problems created by a single replication as we proceed.

Testing these factors with four experimental runs is possible using what Taguchi calls "orthogonal arrays." Orthogonal arrays are mathematically balanced fractional factorial designs. In other words, these mathematically designed experiments can find the main effect of each factor with only a fraction of the combinations used in a full-factorial design.

If we look at the L₄ array presented in table 4.1, we can see how orthogonal arrays are utilized. Beneath the array is a drawing of a line and two dots. This is called the linear graph for the array. Each dot represents where a factor can be assigned. A line connecting two dots represents where an interaction between the two corresponding factors can be studied. If no interaction is desired, then a factor can be assigned to a line. This must be done with caution because, in our example, factor C will now be confounded with any A×B interaction effect.

Table 4.1: The L$_4$ Orthogonal Array.

		Factors		
		A	**B**	**C**
	Col.	1	2	3
Run				
1		1	1	1
2		1	2	2
3		2	1	2
4		2	2	1

Linear Graph

Let's try this method with our three factors and the L$_4$ array. We can assign factor A to column 1, factor B to column 2, and factor C to column 3. Notice how each column contains a series of "1s" and "2s"? These represent the low and high settings, respectively, for the experiment.

We have now completed the design of a Taguchi experiment. To use the experimental design, we follow the rows for each experimental run. First, we randomly select which row to run. For the sake of discussion, let's assume that we have chosen to run row three. This row indicates that factors A and C should be at a high setting and factor B at a low setting.

First run (using row three):

    Factor A = 140° F
    Factor B = 10 minutes
    Factor C = high agitation

With these settings in place, we run a batch of fertilizer and weigh the yield. For our example, we will assume that the yield was 109 pounds. This is recorded on the far right of the array (see table 4.2).

Table 4.2: The completed L₄ array.

		Factors			
		A	**B**	**C**	**Results**
	Col.	**1**	**2**	**3**	
Run					
1		1	1	1	116
2		1	2	2	123
3		2	1	2	109
4		2	2	1	103

Totals by Factors	Factors		
Level	**A**	**B**	**C**
1	239	225	219
2	212	226	232
Grand Total		451	

We repeat this process for the other rows of the design until we have run all four combinations.

Analysis

The analysis of the results of a Taguchi experiment follows the same method we used for full-factorial designs. First, we summarize the results according to the high and low settings for each factor. Factor A serves as an excellent illustration of how this is done.

The first two results (116 and 123) correspond to the low settings for factor A. Added together and divided by two, they will yield the average result for a low *A* setting.

$$\bar{A}_{low} = \frac{116 + 123}{2} = 119.5 \text{ pounds}$$

The last two results (109 and 103) represent the effect of factor A at a high setting.

$$\bar{A}_{high} = \frac{109 + 103}{2} = 106 \text{ pounds}$$

Clearly, changing from a low to a high temperature setting reduced the yield.

We now repeat this procedure for the other factors.

$$\bar{B}_{low} = \frac{116 + 109}{2} = 112.5$$

$$\bar{B}_{high} = \frac{123 + 103}{2} = 113$$

$$\bar{C}_{low} = \frac{116 + 103}{2} = 109.5$$

$$\bar{C}_{high} = \frac{123 + 109}{2} = 116$$

We can now draw effect diagrams using this information. This will give us a rough picture of the effect of each factor (see figure 4.2).

Figure 4.2: The effect diagrams for each factor.

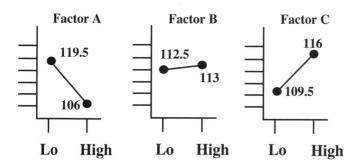

From this simple analysis, we can roughly guess the best settings for the process. However, if we performed an Analysis of Variance (ANOVA), we could learn much more. An ANOVA would tell us how much each factor contributes to the variation in yields; it will differentiate significant from insignificant factors; it will allow us to predict the yield under a combination of settings; and it can even provide us with cost information.

We will find out how to do all of this later. For now, let's concentrate on the initial ANOVA, the significance and contribution of factors. First we calculate the correction factor (CF).

$$CF = \frac{(\Sigma x)^2}{N} = \frac{451^2}{4} = 50{,}850.25$$

Then we calculate the sum of squares for the total experiment.

$$SS_{total} = \Sigma x^2 - CF = 116^2 + 123^2 + 109^2 + 103^2 \\ - 50{,}850.25 = 224.75$$

Next, we calculate a sum of squares for each factor.

$$SSA = \frac{A_{low}^2}{n} + \frac{A_{high}^2}{n} - CF$$

thus,

$$SSA = \frac{239^2}{2} + \frac{212^2}{2} - 50{,}850.25 = 182.25$$

We use a modification of this formula to calculate the effect of the other factors:

$$SSB = \frac{B_{low}^2}{n} + \frac{B_{high}^2}{n} - CF = 0.25$$

$$SSC = \frac{C_{low}^2}{n} + \frac{C_{high}^2}{n} - CF = 42.25$$

We transfer this information to an ANOVA Table (see table 4.3). The degrees of freedom for each factor is one less than the number of levels ($k - 1$). The degrees of freedom for the total experiment is one less than the number of results ($n - 1 = 3$).

Table 4.3: The ANOVA Table for the L₄ array experiment.

Source	df	SS
A	1	182.25
B	1	0.25
C	1	42.25
Total	3	224.75

We determine the variance for each source by dividing the sum of squares by the degrees of freedom. Then we calculate the F-ratios by dividing the variance of each factor by the variance of the error. Unfortunately, there is no error variance in this table. All of the degrees of freedom have been used up by the factors.

Pooling

We select up to half of the factors that seem insignificant and "pool" their sums of squares and degrees of freedom into a "pooled error." In our example, factor B seems to have a very small sum of squares compared to the other factors. Therefore, we pool it into the error (see table 4.4).

Table 4.4: An example of pooling.

Source	df	SS	V
A	1	182.25	182.25
C	1	42.25	42.25
error	1	0.25	0.25
Total	3	224.75	

Caution is necessary when pooling. It is very easy to create a self-fulfilling prophecy by pooling away too many factors. Also, our example would require no pooling if we had performed replications of the experiment. In practice, an experimenter would design an experiment with enough replications that at least six degrees of freedom existed for the error variance.

Please note the very small amount of error detected in this experiment. In practice you will discover that larger experiments with replications will usually have larger experimental errors.

Now we can complete the F-ratio calculations. For example, the F-ratio for the first factor (A) is

$$F = \frac{V_A}{V_{error}} = \frac{182.25}{0.25} = 729$$

Factor C's F-ratio is

$$F = \frac{V_C}{V_{error}} = \frac{42.25}{0.25} = 169$$

The resulting ANOVA Table appears in table 4.5.

Table 4.5: The ANOVA Table with F-ratios added.

Source	df	SS	V	F
A	1	182.25	182.25	729
C	1	42.25	42.25	169
error	1	0.25	0.25	
Total	3	224.75		

At this point, we could go to a table of critical F-values and see if these results are statistically significant. In fact, they are significant for both factors A and C. However, Taguchi presses on in the analysis and looks for the percent contribution each factor makes to the variations in the yield.

To do this, the first step is to create another column for S-prime (S'). This is the intermediate step for calculating the percent contribution (also called %rho).

$$S'_A = SSA - V_{error} = 182.25 - 0.25 = 182$$

$$S'_B = SSB - V_{error} = 0.25 - 0.25 = 0$$

$$S'_C = SSC - V_{error} = 42.25 - 0.25 = 42$$

The S' for the error is the sum of squares for the pooled error plus the variance of errors taken out of each unpooled factor.

$$S'_{error} = SS_{error} + (2 \times V_{error}) = 0.25 + (2 \times 0.25) = 0.75$$

Adding together the products of all of the above calculations yields the S' total for the experiment.

$$S'_{total} = 182 + 42 + 0.75 = 224.75$$

This is used to calculate the percent contribution of each factor.

$$\%rho_A = \frac{S'_A}{S'_{total}} \times 100 = \frac{182}{224.75} \times 100 = 80.98\%$$

Repeating this calculation for factor C and for the pooled error creates the information in table 4.6.

Table 4.6: An ANOVA Table with percent contributions.

Source	df	SS	V	F	S'	%rho
A	1	182.25	182.25	729	182.00	80.98%
C	1	42.25	42.25	169	42.00	18.16%
error	1	0.25	0.25		0.75	0.33
Total	3	224.75			224.75	

Interpretation and use

Our example limits our ability to interpret the results because its lack of replication provides a weak estimate of experimental error. We would undertake a great risk if we used this experiment to make changes in the production process. We simply do not know whether outside factors have been accounted for.

Therefore, we will interpret these results as valid, even though they probably are not. This will demonstrate the interpretation and application of information from an experiment. In practice, we would also want to run confirming experiments.

Clearly, factor A has the greatest effect on yield. Therefore, we refer to our calculated averages and pick the low temperature of 100°. Factor C is also important, so we pick the high setting of high agitation. These two choices create the highest yield. However, factor B contributes very little to the variations in yield. Therefore, we pick a setting for factor B that will allow the greatest economy. For example, the low setting of a 10-minute rest would allow us to increase the speed of production.

In combination, these settings should give us the maximum yield at the lowest cost.

Temperature = 100°

Resting Time = 10 minutes

Agitation = high

As an additional step in the interpretation of the results, we also look at the percent contribution from error. As we will see later, we can learn much about the success of an experiment from the amount and type of error produced. In our example, the error contributes very little to variation. Had the error been high, this could indicate that we missed an important factor in our experiment, or that the process was not under statistical control.

More factors, more information

The first example that we used is far simpler than most experiments conducted in the industrial environment. It also doesn't take advantage of several options you have when experimenting. For example, we made no attempt to study interactions, repeat experimental settings, take multiple samples, or test for various types of error. Therefore, the following examples are presented to introduce you to all of these concepts and practices.

We will begin with the use of more factors, error columns, replications, and testing for interactions. In this example, we will study the effect of five factors. Let's assume that a plastic injection molding engineer is testing the settings on a new molding machine. He has selected a two-level design because of its ease of use and its ability to detect general effects among factors.

The six factors being tested are:

Factor A = injection speed (1 = 3 seconds, 2 = 5 seconds.)

Factor B = mold temperature (1 = 320°, 2 = 360° F)

Factor C = percent regrind (1 = 0%, 2 = 50%)

Factor D = material color (1 = black, 2 = maroon)

Factor E = back pressure (1 = 100, 2 = 200 psi)

In addition to these five factors, the engineer strongly suspects that an interaction occurs between factors A and B. Thus, an A×B interaction is also included in the design.

We select an L$_8$ array because it can test up to seven factors at two levels each. In our case, we place the five factors from the plastic injection machine into the five columns recommended by the linear graph (see figure 4.3). We place the interaction for A×B in column 3.

The seventh column is left open under the designation E. This is shorthand for "error column." Columns 3 and 7 in this design will be analyzed as factors. However, when we conduct the design, we use only the columns for true factors for the experimental run combinations (columns 1, 2, 4, 5, and 6).

Figure 4.3: Assignment of factors and interactions.

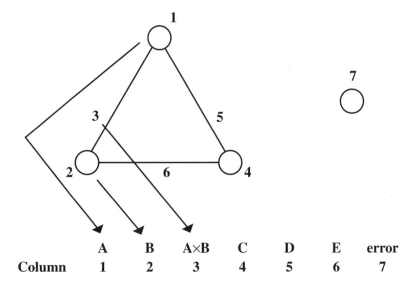

	A	B	A×B	C	D	E	error
Column	1	2	3	4	5	6	7

The error column represents all of the factors that were not included in this study. If we choose our factors correctly, this column will have very little effect on the result because our tested factors represent the significant effects. In analysis, the error column is treated as a separate factor. If it is significantly high, then we know that we failed to include a significant factor in our design.

How to test for interactions

Disastrous results can occur when a real interaction is not included in a Taguchi experiment. An interaction occurring in the experiment will confound other factors and ruin your analysis.

In fact, Taguchi has developed several ways to prevent a significant interaction from being excluded from an experiment. One of the easiest methods is to test suspected interactions in a small experiment. Then obvious interactions are either removed as part of the initial design of the experiment or tested using traditional methods.

In our experiment, factors A and B are assigned to columns 1 and 2. The suspected A×B interaction is assigned to the line connecting the two dots. Thus, A×B is assigned to column 3. During analysis, column 3 is evaluated in the same manner as any factor column.

Repetition and replication

So far, our examples have had only one piece of data for each experimental run. Good experimenters would collect several samples for each experimental run. These multiple samples can tell us more about the experimental errors we measure.

In review: repetitions occur when several samples are taken during a single run. Taguchi calls the error in repetitions secondary error (e_2). Replications occur when we repeat the experimental runs. Taguchi calls this source of error primary error (e_1).

Should a source of experimental error prove significant in an experiment, we will want to know the cause. For example, a high primary error (e_1) means that when we reset the machine to an experimental combination, the accuracy of the machine settings was off or a source of experimental error is present. If the secondary error (e_2) is high, then we know that our samples had too much variation.

When using repetitions, we test the variance of primary error divided by the variance of secondary error. If the F-ratio is significant, the primary error measures the significance of each factor. Otherwise, we pool the two types of error.

$$F = \frac{V_{e_1}}{V_{e_2}}$$

The calculation of factors is the same, but additional sums of squares must be produced.

$$CF = \frac{(\Sigma x)^2}{N} = \frac{374^2}{16} = 8{,}742.25$$

Table 4.7: An L$_8$ array with replications.

Factors

	A	B	A×B	C	D	E	e	Results	
Col.	1	2	3	4	5	6	7		
Runs									
1	1	1	1	1	1	1	1	20	22
2	1	1	1	2	2	2	2	21	21
3	1	2	2	1	1	2	2	23	26
4	1	2	2	2	2	1	1	21	19
5	2	1	2	1	2	1	2	29	26
6	2	1	2	2	1	2	1	31	28
7	2	2	1	1	2	2	1	20	22
8	2	2	1	2	1	1	2	24	21

Level Sum Table

Factors

Level	A	B	A×B	C	D	E	e
1	173	198	171	188	195	182	183
2	201	176	203	186	179	192	191

Note that the sample size has increased to 16 because we have two repetitions of the eight runs.

$$SS_{total} = \sum x^2 - CF = 8{,}936 - 8{,}742.25 = 193.75$$

and,

$$SSA = \frac{A_{low}^2}{n} + \frac{A_{high}^2}{n} - CF$$

or,

$$SSA = \frac{173^2}{8} + \frac{201^2}{8} - 8,742.25 = 49.0$$

...and so on.

Note again that the n factor is eight because of repetitions.

Now we have to break the results into two types of totals for the experiment: the sum of the primaries and the sum of the secondaries. These are represented as

$$SS_{total_1} = \frac{1}{r} (Cell_1^2 + Cell_2^2 + \ldots + Cell_n^2) - CF$$

thus,

$$SS_{total_1} = \frac{1}{2} (42^2 + 42^2 + \ldots + 45^2) - 8,742.25 = 169.75$$

and,

$$SS_{total_2} = \Sigma x^2 - CF = (20^2 + 22^2 + \ldots + 21^2) - 8,742.25 = 193.75$$

The sum of squares for the repetition error is calculated as

$$SS_{error_2} = SS_{total_2} - SS_{total_1}$$

or,

$$SS_{error_2} = 193.75 - 169.75 = 24$$

For the degrees of freedom,

$$df_{error_2} = df_{total} - df_A - df_B - df_C - df_D - df_E - df_{A \times B} - df_{error}$$

79

We calculate the primary error from any empty columns that may remain in the experimental design.

$$SS_{error_1} = \frac{error_1{}^2}{n} + \frac{error_2{}^2}{n} - CF$$

The resulting ANOVA Table appears in table 4.8.

Table 4.8: The ANOVA Table for the L_8 array experiment.

Source	df	SS	V	F	% rho
A	1	49.00	49.00	16.33	23.74
B	1	30.25	30.25	10.08	14.06
A×B	1	64.00	64.00	21.33	31.48
C	1	0.25	0.25	0.08	−1.42
D	1	16.00	16.00	5.33	6.71
E	1	6.25	6.25	2.08	1.68
error (e_1)	1	4.00	4.00	1.33	0.52
e_2	8	24.00	3.00		23.23
Total	15	193.75	12.917		

As you can see, the results show a high error for repetition. About 23 percent of the variation in our results came from the repetitions. In fact, the e_2 error is significantly higher than e_1; therefore we have used the e_2 variance to calculate the F-ratios. This leaves only the A×B interaction as a truly significant factor.

A good experiment would include both repetitions and replications. Along with error columns to seek out missing variables, this type of design would give an experimenter complete understanding of each of the possible sources of error in an experiment. However, maintaining accuracy during the hand calculation of the ANOVA would become very difficult. Thus, computer software is recommended.

Interpretation and use

As the ANOVA results show, the A×B interaction is the strongest effect detected. Therefore, we would want to study the effect of injection speed and mold temperature in combination. Response surface analysis or evolutionary operation are two possible choices for optimizing the interaction. We can set the other tested factors to their most economical levels.

This experiment also shows a typical amount of error occurring. This represents the sensitivity of the experiment. With only two replications, we still had a good deal of background "noise" in this experiment. Still, we detected some real effects, and the noise did not indicate that we may have missed an important factor.

Experiments at three levels

Two-level experiments are great to use as screening experiments and simple investigations. However, many industrial processes will need information that can be directly translated into engineering data. Experiments with factors set to three or more levels can accomplish this task.

The next example will illustrate the importance of the information obtained. Let's assume that an engineer is testing a new control circuit for a microwave oven. In operation, 45 volts DC powers the circuit. The circuit then sends about 5 volts DC to various control units.

The engineer is curious about the effect of voltage variations going into the circuit and the amount of control which manufacturing can induce by adjusting two variable resistors in the circuit. Thus, the following factors are tested at three levels each.

Input voltage (40, 45, and 50 volts)
Resistor 1 (100, 110, and 120 ohms)
Resistor 2 (45, 50, and 55 ohms)

As a response, the engineer will measure the volts DC that the circuit produces. He is hoping that the experiment will tell him how to design the circuit to produce the most consistent output voltage.

We place the three factors in an L9 array (see table 4.9). This array can test up to four factors at three levels each. In our example, we leave the fourth column open as an error column.

Table 4.9: The L9 orthogonal array.

Factors

		A	B	C	e	
	Col.	1	2	3	4	
Runs						Results
1		1	1	1	1	5.5
2		1	2	2	2	5.3
3		1	3	3	3	5.9
4		2	1	2	3	4.4
5		2	2	3	1	4.7
6		2	3	1	2	5.0
7		3	1	3	2	4.9
8		3	2	1	3	5.6
9		3	3	2	1	5.7

As you can see from the raw data, the actual output voltages varied widely. However, the ANOVA will determine if this effect was created by a particular factor.

The ANOVA proceeds in a fashion very similar to that of the L4 and L8 arrays.

$$CF = \frac{(\Sigma x)^2}{N} = \frac{47^2}{9} = 245.44$$

$$SS_{total} = \Sigma x^2 - CF = (5.5^2 + 5.3^2 + \ldots + 5.7^2)$$
$$- 245.44 = 2.02$$

Each factor's sum of squares is calculated using the formula

$$SS_{factor} = \frac{A_1^2 + A_2^2 + A_3^2}{n} - CF$$

where

A_1 = the total of the low settings

A_2 = the total of the medium settings

A_3 = the total of the high settings

n = the number of numbers in a single level's group

Thus,

$$SSA = \frac{16.7^2 + 14.1^2 + 16.2^2}{3} - 245.44 = 1.27$$

therefore,

$SSB = 0.54$

and,

$SSC = 0.096$

and,

$SSe = 0.109$

Please note that the degrees of freedom for factors at three levels now becomes two ($df = k - 1$, or $df = 3 - 1 = 2$).

We then place these results into an ANOVA Table (see table 4.10).

Table 4.10: The ANOVA Table for the L₉ array experiment.

Source	df	SS	V
A	2	1.269	0.634
B	2	0.542	0.271
C	2	0.096	0.048
error	2	0.109	0.054
Total	8	2.016	0.252

By pooling out the insignificant factors, *C* and *Error*, we can complete the calculation (see table 4.11).

Table 4.11: The ANOVA Table after pooling.

Source	df	SS	V	F
A	2	1.269	0.634	12.37
B	2	0.542	0.271	5.29
C	2	0.096	0.048	
error	2	0.109	0.054	
Total	8	2.016	0.252	
pooled error	4	0.205	0.051	

You could proceed as usual to select the best settings for all factors, since the pooled error is not too large. However, look at the degrees of freedom for each factor. Because we tested at three levels, we have two degrees of freedom for each factor. Wherever there is more than one degree of freedom, we can decompose our results into more information.

With two-level designs, the response of each factor could only be drawn as a straight line. However, having three levels allows us to look for curves in the response. A curve represents a dynamic response to a factor, not a linear response. A dynamic response presents many possibilities for engineering a better design.

The three-level responses can be decomposed into linear and quadratic effects. To do this, we expand the ANOVA Table. The formula for the linear effect is

$$SS_{lin} = \frac{((-1 \times A_1) + (0 \times A_2) + (1 \times A_3))^2}{r \times \lambda^2 s}$$

where

r = the number of repetitions (3 in this example)

$\lambda^2 s$ = a factor found in Appendix C

and the quadratic effect is,

$$SS_{quad} = \frac{((1 \times A_1) + (-2 \times A_2) + (1 \times A_3))^2}{r \times \lambda^2 s}$$

To obtain the factors for the formulas, refer to the table in Appendix C. Doing so for our example yields the following results.

$$SSA_{lin} = \frac{(-16.7 + 16.2)^2}{3 \times 2} = 0.042$$

$$SSA_{quad} = \frac{(16.7 - (2 \times 14.1) + 16.2)^2}{3 \times 6} = 1.227$$

and,

$$SSB_{lin} = \frac{(-14.8 + 16.6)^2}{3 \times 2} = 0.54$$

$$SSB_{quad} = \frac{(14.87 - (2 \times 15.6) + 16.6)^2}{3 \times 6} = 0.002$$

85

Adding these to the ANOVA Table results in the following.

Table 4.12: The ANOVA decomposed into linear and quadratic effects.

Source	df	SS	V	F
A	2	1.269	0.634	
B	2	0.542	0.271	
C	2	0.096	0.048	2.18
error	2	0.109	0.054	2.46
A_{lin}	1	0.042		
A_{quad}	1	1.227	1.227	55.77
B_{lin}	1	0.540	0.540	24.54
B_{quad}	1	0.002		
Total	8	2.016	0.252	
pooled error	2	0.044	0.022	

From this we can see that factor A is significant as a quadratic and factor B has a linear response. To illustrate graphically what this means, we create effect diagrams. The responses at each level for the two factors are:

$$A_1 = 5.567$$
$$A_2 = 4.700$$
$$A_3 = 5.400$$

$$B_1 = 4.933$$
$$B_2 = 5.200$$
$$B_3 = 5.533$$

Drawn as an effect diagram, factor A would clearly create a quadratic effect (see figure 4.4). A setting at A_2 would then minimize variation in the output voltage. By adjusting the circuit design, the engineer can raise the bottom of this curve to the target of 5 volts DC.

Figure 4.4: Effect diagrams for three–level factors.

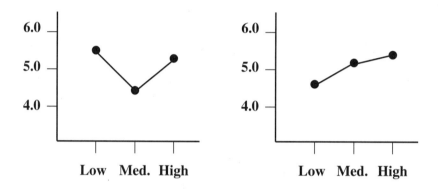

Modifying arrays to fit your needs

Needless to say, the world will not always provide you with neat situations where all of the factors are either at two or three levels. Mixtures of levels and levels of four or more commonly occur. To cope with these situations, we can modify the orthogonal arrays.

One factor at four levels

The first example of such a modification is to create an array in which we test one factor at four levels. An L_4 array quickly illustrates this modification.

First, we select three columns from a linear graph. The three must be two dots and the line that connects them. Since a four-level factor has three degrees of freedom, we need three columns with one degree of freedom each.

Next, we take the first two columns, in this case columns 1 and 2. The combinations they create are

1,   1

1,   2

2,   1

2,   2

These four unique combinations will represent the four levels of the array. The finished product looks like the array in table 4.13.

Table 4.13: Finished four-level factor column.

		Factors		
Run	A	B	~~B~~	~~C~~
1	1			
2	2			
3	3			
4	4			

We determine the effect of the four-level factor by using the same formula used for other sums of squares.

$$SSA = \frac{A_1{}^2}{n} + \frac{A_2{}^2}{n} + \frac{A_3{}^2}{n} + \frac{A_4{}^2}{n} - CF$$

The advantage of a four-level factor is that you can test a wider range of responses and find more complex response curves. For example, a four-level factor can be tested for linear, quadratic, and cubic effects.

Mixing levels within an array

In many experiments it will be impossible to test all factors at, say, three levels. A good example is a machining process that has only two types of steel that can be used. A third level is impossible to test.

The solution is to use a three-level factor array and to add a third "dummy treatment" to the two-level factor. To see how this works, let's assume that you wish to test two factors at three levels and one factor at two levels.

$$A_1 \quad A_2$$

$$B_1 \quad B_2 \quad B_3$$

$$C_1 \quad C_2 \quad C_3$$

We use an L_9 array to assign factors B and C to the first two columns. We assign the two-level factor to the third column. Then the 3s in the column become the dummy treatments. Where a 3 is present in the third column, one of the two levels is repeated. (See table 4.14)

The formula for the sum of squares is altered a little to account for this modification.

$$SSA = \frac{(A_1 + A')^2}{n \times r} = \frac{A_2^2}{n} - CF$$

where

n = the number of numbers in a group
r = the number of repeated runs

Although this is an effective way to modify an array, it should not be used to mix several two-level factors with several three-level factors. In addition, there are some unique arrays (e.g., L_{18}) that have been designed from scratch to include a two-level factor with three-level factors.

The Experimenter's Companion

Table 4.14: Using a dummy variable in an L9 array.

		Factors			
	Col.	A 1	B 2	C 3	e 4
Run					
1		1	1	1	1
2		1	2	2	2
3		1	3	1′	3
4		2	1	2	3
5		2	2	1′	1
6		1	3	1	2
7		3	1	1′	2
8		3	2	1	3
9		3	3	2	1

Parameter design

One of the most common applications of the Taguchi approach is the optimization of a process. This is called parameter design because the parameters of a process are under study. However, this method also uses one of the most powerful techniques Taguchi ever developed, the application of an outer array.

To explain the entire method, let's return to an example we used earlier. A chemical batch mixing experiment used an L4 array to test three controllable factors.

A = temperature (100°, 140° F)
B = time (10 minutes, 20 minutes of mixing)
C = agitation (low speed, high speed)

Missing from this experiment were factors that are not so easy to control. The production department identified three that they believed play havoc with output quality. These "outer noise" factors are identified with a prime symbol (′) in the factor designation.

A′ = humidity (low = 50%, high = 95%)
B′ = ambient temperature (low = 65°, high = 90°)
C′ = barometric pressure (low = 29 inches, high = 31 inches)

According to Dr. Taguchi's philosophy, we should try to make the process robust against these uncontrollable factors. In other words, we should find the proper settings for the machine that will make the quality of the product consistent no matter what the humidity, temperature in the plant, or barometric pressure.

So far we ran an L4 array to find the best settings on the machine. However, that experiment did not take into account these uncontrollable factors. Therefore, we shall run a second experiment in which the uncontrollable factors are set into a separate array called the "outer array." The inner array will be the original L4 design for the controllable factors.

Figure 4.5 demonstrates the appearance of this design.

Figure 4.5: A parameter design.

	Run		Outer Array			
	A′	1	1	1	2	2
	B′	2	1	2	1	2
	C′	3	1	2	2	1

Inner Array

Run	A	B	C
1	1	1	1
2	1	2	2
3	2	1	2
4	2	2	1

Data Area

The way to conduct this type of experiment is to replicate the inner array four times. On each replication you must either induce or wait for the conditions the outer array describes for each of its runs. For example, on the first replication humidity, ambient temperature, and barometric pressure are at their low settings.

low humidity (about 50% relative)

65°

29 inches of mercury

We conduct the four experimental runs of the inner array under these conditions. Then we set the conditions to those specified in the second column of the outer array and replicate the four inner array runs. This continues until all outer array replications are completed.

Analysis

The data gathered for this experiment is the pounds of fertilizer the process yielded. The raw data are presented below.

Table 4.15: Inner and outer arrays and data.

		Outer Array			
	Run	1	2	3	4
	1	109	110	108	106
	2	112	114	107	110
Inner	3	108	112	106	109
Array	4	116	111	113	107

The analysis involves an ANOVA for the inner array and another ANOVA for the outer array. These will be combined to form a single ANOVA Table.

Factors A, B, and C are used to calculate sums of squares just as before. However, this time there are replications to give us a better picture of the sources of error. The sum of squares for factor A is calculated as,

$$CF - \frac{(\Sigma x)^2}{n}$$

or,

$$CF = \frac{1758^2}{16} = 193,160.25$$

Each factor's effect is calculated using the standard formula

$$SSA = \frac{A_1^2}{n} + \frac{A_2^2}{n} - CF$$

For our example, this would form the following equation.

$$SSA = \frac{(109 + 110 + \ldots + 110)^2}{8} + \frac{(108 + 112 + \ldots + 107)^2}{8}$$
$$- 193,160.25 = 2.25$$

This is repeated for the other factors. Then the sums of squares for the outer array factors, A', B', and C', are calculated. The same formula is used, but the data are gathered by going up and down the columns instead of across the rows.

$$SSA' = \frac{(109 + 112 + \ldots + 111)^2}{8} + \frac{(108 + 107 + \ldots + 107)^2}{8}$$
$$- 193,160.25 = 42.25$$

Also note that we have 16 pieces of data and 15 degrees of freedom. Each of the six factors involved has one degree of freedom. This leaves nine degrees of freedom for the error. In this case, this is primary error. The resulting ANOVA Table is outlined in table 4.16.

Table 4.16: ANOVA with calculation of primary error.

Source	df	SS	V	F	% rho
A	1	2.25	2.25	0.491	–1.80
B	1	30.25	30.25	6.60	19.78
C	1	0.25	0.25		
A′	1	42.25	42.25	9.22	29.03
B′	1	0.00			
C′	1	1.00			
e_1	9	53.75	5.97		
pooled	12	55.00	4.58		52.99
Total	15	129.75	8.65		

As you can see, factor B (time) is the most significant factor to set in this process. However, the outside factor A' (humidity) is also a significant factor. Therefore, to pick the proper settings, factor B must be compared to A'. The best setting will be the one that creates the highest yield for both high- and low-humidity conditions. In other words, we are seeking a setting that will make the yield robust against the effects of humidity.

The best method for doing this is to calculate the averages for high and low time against high and low humidity.

$$\overline{B_1 A'_1} = \frac{B_1 A'_1}{n} = \frac{109 + 110 + 108 + 112}{4} = 109.75$$

$$\overline{B_1 A'_2} = \frac{B_1 A'_2}{n} = \frac{108 + 106 + 106 + 109}{4} = 107.25$$

$$\overline{B_2 A'_1} = \frac{B_2 A'_1}{n} = \frac{112 + 114 + 116 + 111}{4} = 113.25$$

$$\overline{B_2 A'_2} = \frac{B_2 A'_2}{n} = \frac{107 + 110 + 113 + 107}{4} = 109.25$$

The resulting table of averages is

	Humidity	
Factor B	Low	High
Low	109.75	107.25
High	113.25	109.25

The "High" time setting creates the largest yields in both humidity settings. Therefore, B_2 is a selected setting. Since the other inner factors are not significant, we can set them to the most economical levels.

Summary

We have seen how using an outer array has helped us design a process that is resistant to the normal causes of variation. This is a powerful tool for increasing the quality of a process while finding the least expensive settings that maintain this quality. The result is usually high quality at a low cost.

Tolerance design

Only after using parameter design and other experimental tests to optimize a process should you consider setting tolerances for maximum quality and economy. This process is called tolerance design, and it usually involves the greatest difficulties and expenses in making changes. An example will illustrate why.

In our fertilizer mixing experiment, we assumed that the chemical nature of the raw materials was consistent. In reality these will change from lot to lot. At extra expense, the company could buy more consistent raw materials or adjust the materials before processing. However, this extra money would be spent in ignorance.

At this point, the company has no idea which chemical characteristics of the material are important to the results. A tolerance design experiment could reveal which of the materials need tighter tolerances and which tolerances can be loosened to allow the purchase of less expensive materials.

As an illustration, let's take the example of two chemicals, ammonia and phosphate. Three characteristics of the ammonia base and the phosphate compounds are

pH
percent moisture
percent phosphate

Records indicate that these three characteristics vary. Below are the averages and standard deviations observed.

Characteristic	Average	Standard Deviation
pH	8.5	0.5
% moisture	3%	0.25%
% phosphate	5%	0.75%

To test the effect of these variations on the final yield of the now-optimized process, we construct an L9 array. We determine each of the settings by taking the average value as a middle level and the square root of one-and-a-half times the standard deviation as each of the other settings.

$$\text{High level} = \bar{x} + \sqrt{\frac{3}{2}}\ \sigma$$

$$\text{Optimal} = \bar{x}$$

$$\text{Low level} = \bar{x} - \sqrt{\frac{3}{2}}\ \sigma$$

Thus, the final levels for the L9 array are:

pH (1 = 7.9, 2 = 8.5, 3 = 9.1)

% Moisture (1 = 2.7%, 2 = 3.0%, 3 = 3.3%)

% Phosphate (1 = 4.1%, 2 = 5.0%, 3 = 5.9%)

We conduct the experiment and generate the following ANOVA Table (see table 4.17).

Table 4.17: ANOVA Table with decomposition.

Source	df	SS
A	2	34.89
B	2	284.22
C	2	10.89
error	2	2.89
A_{lin}	1	0.17
A_{quad}	1	34.72
B_{lin}	1	240.67
B_{quad}	1	43.56
C_{lin}	1	10.67
C_{quad}	1	0.22
Total	8	332.89

Because each factor was decomposed into linear and quadratic effects, we can quickly determine which factors require tight tolerances. For example, factors B and C (% Moisture and % Phosphate) have a linear effect on the yield. Therefore, tight tolerances will be necessary to control variations in results.

Factor A has more of a quadratic effect. This means that we can find an optimal point on the response curve. This helps us to set both an ideal tolerance and a wider range of variation. Thus, we have found that pH can vary considerably more than we expected. This allows us to purchase from more vendors or at a lower cost.

Summary

Tolerance design is the last step in a Taguchi process; you should conduct all other sources of improving a process first. For example, test and optimize the actual design of a product before setting the tolerances.

Signal-to-noise ratios

The signal-to-noise ratio is one of Taguchi's most famous but controversial additions to the science of experimenting. The signal-to-noise ratio is expressed exactly the same as the formula you see listed on audio equipment at a department store—in decibels. It measures the ratio between the signal generated and the background noise. The higher the ratio, the better the sound quality.

Taguchi applies this universal method of measuring communication efficiency to experimentation in a unique fashion. He translates the raw data from the experimental results into signal-to-noise ratios. In other words, he makes the signal-to-noise ratio a new response factor, or as he calls it, a "quality characteristic."

What makes this application unique is that the signal-to-noise (S/N) ratios have been designed to increase as the settings for factors change to create the least amount of error variance. That is, a high S/N indicates optimal settings for a process.

The controversy surrounding the use of S/N ratios is that they tend to suppress the experimental variation in the analysis. Experiments with experimental error above 50 percent, which should have been re-run, have shown strong factor responses when analyzed as S/N ratios.

Calculating S/N

The calculation of S/N ratios and their analysis results in an ANOVA Table. However, creating such a table requires a few preliminary steps. The first is to determine which of the following conditions your experiment addresses.

1. **The nominal is the best**–You have a specific target figure to achieve, such as the optimal tolerance.

2. **The smaller the better**–You desire lower values, such as lower standard deviations.

3. **The larger the better**–Higher values are more important, such as increased strength in the final product.

Once you have selected the condition, calculate the S/N ratio using the appropriate formula which follows.

Nominal is best

$$\text{Variance} = \frac{(X_1 - \bar{x})^2 + (X_2 - \bar{x})^2 + ... + (X_n - \bar{x})^2}{n}$$

Smaller is better

$$\text{Variance} = \frac{X_1^2 + X_2^2 + ... + X_n^2}{n}$$

Larger is better

$$\text{Variance} = \frac{1}{X_1^2} + \frac{1}{X_2^2} + ... + \frac{1}{X_n^2}$$

Let's assume that we used an L$_4$ array in an experiment and obtained the following results. (see table 4.18).

Table 4.18: Data for calculating a signal-to-noise ratio.

		Factors				
		A	**B**	**C**		
	Col.	**1**	**2**	**3**		
Runs					**Results**	
1		1	1	1	20	19	18
2		1	2	2	22	21	24
3		2	1	2	19	20	17
4		2	2	1	21	20	19

In this situation, the experimenter is testing pieces of rubber with a durometer. The objective is to produce rubber pieces with a durometer reading of 20. This would make this example a "Nominal is Best" situation.

In the first run of data the variance is

$$\text{run}_1 = \frac{(20-20)^2 + (20-19)^2 + (20-18)^2}{3} = 1.67$$

Thus, the signal-to-noise ratio becomes

$$S/N = -10 \log (1.67) = -2.23 \text{ db}$$

We repeat this for each experimental run. Therefore, the final S/N translations are

$\text{run}_1 = -2.23 \text{ db}$
$\text{run}_2 = -8.45 \text{ db}$
$\text{run}_3 = -5.22 \text{ db}$
$\text{run}_4 = +1.74 \text{ db}$

Adding these to the L_4 array creates the response table needed to complete an ANOVA analysis (see table 4.19).

Table 4.19: Signal-to-noise ratios for the L_4 experiment.

		Factors			
		A	**B**	**C**	
	Col.	1	2	3	**S/N**
Run					
1		1	1	1	−2.23
2		1	2	2	−8.45
3		2	1	2	−5.22
4		2	2	1	+1.74

We treat the S/N numbers like any other response variable and conduct the standard ANOVA analysis (see table 4.20).

Table 4.20: ANOVA Table for signal-to-noise ratios.

Source	*df*	*SS*	V	F	%rho
A	1	12.970	12.97	90.67	22.57
B	1	0.143			
C	1	43.710	43.71	305.64	76.68
Total	3	56.520			
pooled	1	0.143	0.76		

As you can quickly see, factor B has been pooled out and factors A and C create the most significant effects. Looking back to our S/N totals for each level, we can see that the A_2 / B_2 / C_1 settings would be most effective.

Level Summary
A_1 = –10.68 db
A_2 = –3.48 db

B_1 = –7.45 db
B_2 = –6.71 db

C_1 = –0.49 db
C_2 = –13.67 db

To measure the increase in performance from these settings, Taguchi uses "gain." Gain is expressed as the difference between a previous S/N ratio and a new S/N ratio. For example, the gain from switching from A_1 to A_2 is

Gain = S/N of new setting – S/N of old setting

The new settings have S/N ratios of –3.48, –6.71 and –0.49. Together, they add up to –10.68. The old settings equal the total S/N for the experiment, or –14.16. Thus, the gain is

Gain = –10.68 – (–14.16) = 3.48 db

This provides you with a relative measure of improvement.

Taguchi's loss-function

A large section of Taguchi's work uses what he calls a loss-function. The loss-function defines the loss to society when a product has not been made to optimal specifications. In other words, the loss-function expresses what we in the West would call "process capability" in economic terms.

Although Taguchi uses the loss-function to make many decisions about plant operations and product designs, the U.S. has failed to adopt it as a valid measure of quality. Therefore, we will only briefly review how it works so that you can be more familiar with its use in Taguchi's writings.

Taguchi rejects the American view that the tolerance on a specification represents a point where a product is suddenly "bad." Instead, he sees the optimal specification as the target all products should achieve. As products deviate from this target, the loss to society increases exponentially. The resulting quadratic curve of loss is called the loss-function (see figure 4.6).

Figure 4.6: The Taguchi loss-function.

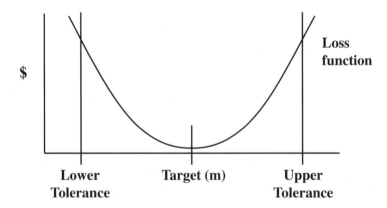

Calculating a loss-function is fairly easy and straightforward. As mentioned in the previous section, one of three conditions occurs relative to specifications. Either:

1. you are trying to achieve an optimal specification,

    or

2. the smaller the characteristic, the better (such as the number of defects),

    or

3. the larger the characteristic, the better (such as pull strength).

What follows is an example in which a specific target is to be achieved. The formula is

$$L(y) = k [\sigma^2 + (\bar{y} - m)^2]$$

where

$$k = \frac{A_o}{\Delta^2}$$

and

A_o = the cost of a countermeasure, usually the cost to scrap a part

Δ^2 = the distance from the target to a tolerance limit, squared

$L(y)$ = the loss function

$\bar{y}$ = the average part size

m = the target, or optimal specification

σ = standard deviation

An example can quickly illustrate the calculation. We are assuming that a capacitor is supposed to be 500 μ. The tolerance for this part is plus or minus 5 μ. If the part is out of tolerance, it costs $6 to scrap the circuit board.

A capability study has found an average strength of 502 μ and a standard deviation of 1.5 μ. The loss-function calculation begins with the calculation of k.

$$k = \frac{A_o}{\Delta^2}$$

or

$$k = \frac{\$6}{5^2} = \$0.24$$

The rest of the calculation now falls into place.

$$L(y) = \$0.24\ [1.5^2 + (502 - 500)^2]$$

$$L(y) = \$1.50$$

By multiplying this loss times the number of circuit boards being produced, we can estimate the loss to society from both the variation in production and the capacitor's strength being 2 μ off target.

Other formulas

When "smaller is better" characteristics are involved, the formula used is

$$L = k\,(\sigma^2 + \bar{y}^2)$$

For "larger is better" characteristics, the formula becomes

$$L = A_o\, \Delta_o^2\, \frac{1}{y^2} \left(1 + 3\,\frac{\sigma^2}{\bar{y}^2}\right)$$

In Taguchi's work, every experiment begins and ends with a loss-function calculation. We use the first calculation the same way we would use a capability study to determine existing conditions. We make the second calculation after implementing the new settings. We compare this loss-function to the first to measure the amount of improvement from experimentation.

If you would like to learn more about the loss-function, several of Taguchi's works cover it in detail. In addition, several articles have been written about it in popular trade journals.

Chapter Five
Model Building and
Confirming Experiments

In this section we talk about:
- a simple regression model
- a multiple regression model
- the use of response surfaces
- how and when to run a confirming experiment

Mathematical models

Completing the statistical tests and choosing the settings for a process or design does not end the experiment. From this point you must build a mathematical model of the dynamics you have discovered. You will use this model to predict the results of your suggested settings. Then you must confirm this prediction through a second experiment or by making a production run.

Model building

The techniques for building an accurate mathematical model of an experiment can be difficult for someone unfamiliar with mathematics. The person who constructs the model should be very competent in the use of mathematics. This is not to say that the actual model building has to be difficult or even complicated. However, it is critical that the person building the model be familiar with all of the choices available. Model building itself is a subject for separate and serious study.

A simple model

A very simple mathematical model is the regression. An example can illustrate how this model is easily formed. Let's look at the experiment which studied fertilizer mixing. A simple experiment would be to test for possible correlations between one of the factors and the results (pounds yield of fertilizer).

Let's further suppose that we collected the data in table 5.1.

A simple correlation test would reveal that the temperature of mixing has a correlation coefficient of -0.931 to the yield. This represents a significant correlation. The model that is built from a regression is called the line of regression (see figure 5.1).

$$Y = a + bX$$

Table 5.1: Data from the chemical mixture experiment.

Temperature	Yield
100°	116
100°	123
140°	109
140°	103
100°	121
100°	119
140°	107
140°	101

Figure 5.1: The model of regression.

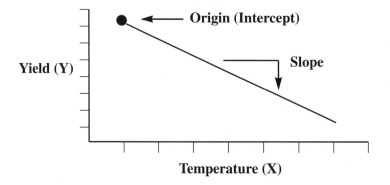

This formula represents a straight line that best fits the data. In plain English, the formula says,

average yield = intercept + (slope × temperature)

In this example, we attempt to predict the amount of yield by specifying a particular temperature of mixing. For example, what would be the average yield for a temperature of 120° during mixing?

To construct the final model, we need to solve for a and b, the intercept and slope of the line. The formulas for these values are based on the raw data we collected during the experiment.

$$b = \frac{n\,(\Sigma XY) - (\Sigma X)\,(\Sigma Y)}{n\,(\Sigma X^2) - (\Sigma X)^2}$$

$$a = \frac{\Sigma Y}{n} - b\,\frac{\Sigma X}{n}$$

where X = the prediction data

Y = the response data

n = the number of data pairs

Notice how we solve b before a.

Solving these equations using the raw data presented above yields the following results.

$b = -0.369$

$a = 156.625$

Placing these values into the equation from the regression line creates our mathematical model.

$Y = 156.625 + (-0.369\ X)$

Inserting a 120° value for X,

$Y = 156.625 + (-0.369 \times 120)$

or,

$Y = 112.345$

Thus, our model predicts an average yield of a little more than 112 pounds of fertilizer.

This is not the end of the model building process. As you know, this is only an average result. The actual results can vary. Therefore, you would continue with the model building process by calculating standard errors for the regression line. Applying these to the formula would create "zones of error" around the regression line. (See figure 5.2.) Thus, you can make a more complete prediction, such as "the yield at 120° should be 112 pounds, plus or minus seven pounds."

Figure 5.2: Zones of error around the line of regression.

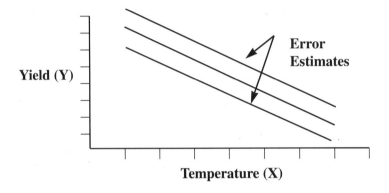

The multiple regression model

As we discussed earlier, most experiments involve the study of several factors at the same time. Therefore, the simple regression model presented above would not apply. Instead, we would calculate a multiple regression and create a larger mathematical model. Usually, such a model takes the following form.

$$Y = a + b_1 X_1 + b_2 X_2 + \ldots + b_n X_n$$

Each b is a slope calculation for each X. The size of your final model will depend on the number of factors to be included.

Constructing a multiple regression model involves making several regression tests.

1. Using your original experimental design and data, correlate each factor to the results. Keep only those factors that have 95 percent significance in their correlations.

2. Check the standard deviation of the remaining factors to ensure that the variations within each factor are consistent.

3. Correlate each remaining factor with the other factors. Remove any factor that correlates with another factor more than 0.70. This will eliminate confounding factors.

4. T-test the surviving combination of factors. Remove one factor at a time and retest until you have the significant factors. This is also called "step-wise regression analysis" in most statistical software.

5. Check for linearity and normality in the remaining data.

6. Calculate the multiple regression.

7. Form the math model.

8. F-test for the significance of the model.

These steps are time-consuming and demanding when done by hand. The good news is that most high-grade statistical packages will do these calculations for you.

An example will illustrate the type of model that is produced. Let's look again at the fertilizer mixing experiment from the previous section.

Table 5.2: The chemical mixture experiment and data from all factors.

Temperature	Time (minutes)	Agitation	Yield
100°	10	1	116
100°	20	2	123
140°	10	2	109
140°	20	1	103

Our ANOVA results told us that factors A and C (temperature and agitation) were significant. A multiple regression analysis would confirm this finding. Thus, "time" would be eliminated from the model. The resulting model, based on a multiple regression, would be

$$Y = 143.5 - 0.388 \text{ (temperature)} + 6.5 \text{ (agitation)}$$

All we need to do is insert values for temperature and agitation to make a prediction. For example, at a temperature of 120° with a high measure of agitation,

$$Y = 143.5 - (0.388 \times 120) + (6.5 \times 2) = 115.94$$

In other words, we could expect a yield of about 116 pounds. And, again, we could calculate standard errors for each factor to obtain a more complete estimate.

Response surface–a non-linear model

If you have ever worked with a topographical map, you will be able to visualize how the response surface method works. You begin with two factors from your experiment. Lay out the factor levels as a grid on a flat piece of paper (see figure 5.3). Plot the response at each grid coordinate. Also, sample additional responses for a point in the center of the grid.

Figure 5.3: Design of a response surface.

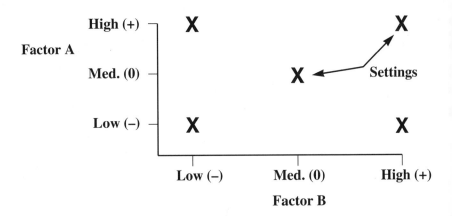

Examining the data from our fertilizer experiment, we would begin by modeling the two significant factors, temperature and agitation. To these we would add three data points taken from the center point of the grid (see table 5.3).

Table 5.3: Experimental design for surface response.

Temperature	Agitation	Results
–	–	116
–	+	123
+	+	109
+	–	103
0	0	112
0	0	114
0	0	107

We lay out these results on the grid of settings. The zeros represent a medium setting for the factors (medium agitation at 120°), and the three results are averaged together. If a peak or valley appears at the center point, this is usually the optimal setting area. We would then run another experiment with more levels to pinpoint the optimal peak or valley.

The following illustration shows that neither a peak nor a valley exists at the center point for our example. (See figure 5.4.) Therefore, we conduct a few more steps to find where such a point exists.

Figure 5.4: The results of response surface analysis.

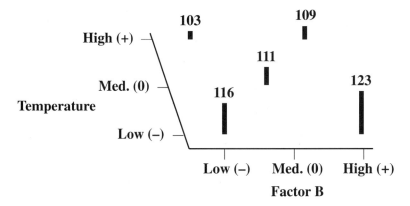

Matching each result to the corresponding "sign" in the design will yield the direction of the peak or valley.

$$\text{Temperature } (b_1) = \frac{1}{4} (-116 - 123 + 109 + 103) = -6.75$$

$$\text{Agitation } (b_2) = \frac{1}{4} (-116 + 123 + 109 - 103) = 3.25$$

What these two results tell us is that the peak we are seeking is in a direction defined as a decrease of 6.75 units of temperature for every increase of 3.25 units of agitation. In other words, the optimal is probably at a lower temperature under greater agitation.

Thus, the experimental design is shifted in this direction and retested. Once we roughly locate the "peak," we conduct a larger experiment to give greater resolution to the results. This helps to find the exact point of optimization.

The response surface method has many more applications than the ones we are describing here. We can find and describe ridges and valleys. We can define the slope of combined factors to locate points where we could make a design more robust. (See figure 5.5.) In short, response surface methods are a class in and of themselves.

Figure 5.5: Shifting the response surface.

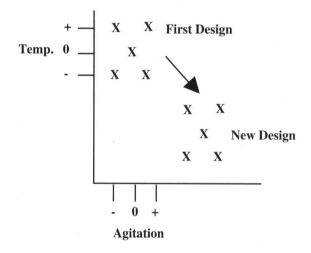

Confirming experimental results

As you may have already gathered from the methods of model building outlined above, there is a pressing need for further experiments to confirm the prediction we may make.

There are two basic methods for confirming your mathematical model. One is to run additional experiments, and the other is to try the suggested settings in a production run. In a few rare instances, you may use computer simulation to confirm your results.

Confirming experiments

In the scientific community, the ideal is to publish your results and wait for another group using similar equipment to re-run your experiment to obtain similar findings. In industry this approach is impractical. Instead, the experimental group will usually repeat the original experiment as confirmation. In some cases, a different group may repeat the experiment to protect against bias.

Sometimes these confirming experiments are smaller than the original experiment. For example, an L_{27} array that tested ten factors at three levels each may have found three factors that are significant. The confirming experiment may then be an L_9 array.

Perhaps an interesting interaction became evident during an initial experiment. The experimenter would then test this interaction further using response surface design and analysis.

Another possibility would be to include the results of several experiments into a single experiment. For example, the Quality department may have several factors it has correlated to improved product quality, while Engineering may have a list of factors that would make the product design more robust. A single, grand experiment could test these factors for interaction and actual effect under production conditions.

Yet another form of confirming experiment is larger than the original experiment. Using the parameter design method, the factors found significant in a preliminary experiment are exposed to several "environmental effects" in an outer array. Each originally suggested setting for the factors is evaluated for robustness against these outside influences common to manufacturing. The settings that hold up well against environmental variation are kept.

Confirming production runs

A simple way to confirm the results of an experiment is to suggest some new settings, predict their results, and then make a pilot production run. If the pilot parts match the prediction, then the validity of the experiment and the mathematical model is supported.

Of course, for a thorough investigation, several alternative settings would also be tested during production to confirm the robustness of the math model at all levels. This will require making some bad parts, but the information gained could be valuable.

However, in most cases, a company will merely apply the suggested settings to the process or design and continue to manufacture the product. The Quality Assurance people will monitor any changes in the product's quality and report these to the experimenters.

After 30 days of production under statistical control, the Cpk ratios obtained would be compared to those predicted. The idea is to confirm the validity of experimental results over time.

Summary

We have seen that there are several alternative methods for forming mathematical models to fit experimental results. We have also seen that becoming well-versed in their use requires a separate course. However, do not let this discourage you from trying to fit models to data. You may test the resulting predictions using confirming experiments, and this validates your original results.

Chapter Six
The Ethics and Presentation
of Experiments

In this section we talk about:
- the ethical threats to experiments
- the countermeasures to those threats
- the oral presentation of an experiment
- the written presentation of an experiment

Ethics

Being an experimenter carries the same responsibilities as being a statistician. Even at the cost of your job, you must protect the integrity of the information you produce. There is a large degree of public trust necessary to make the application of experimental results possible. Take the case of the polio vaccine. Had the public disbelieved the original experiments, polio could still be with us today.

Every company should take steps to assure the accuracy, validity, and integrity of its research. Several departments, and the competitive position of the company itself, will usually depend on correct and usable information from experimentation. Careful preparation and strict rules of research will prevent problems in the effectiveness of the experiment and the ethics of the researcher. Also, chances are that the experimental results will be put to immediate use in production. The results of mistakes and exclusions would be detected at once. Thus, it is in the best interest of the experimenter to keep complete records so that problems in experimental methods can be traced and corrected.

You should maintain an experimenter's log book that records everything you did in an experiment. In addition, the original data you collect should be used in its entirety. If editing is necessary, you should write a memo to your boss explaining what will be edited and why this is justified.

A strongly recommended procedure is peer review. As an experimenter you should actively seek out the opinions of others trained to experiment. Your research proposal, experimental design, and results should all be reviewed. This will help improve your work and add validity and trust to your results.

Finally, publication should be a natural part of your experimental work. Initially, you may only publish short reports about your work for use within the company. Eventually, you will want to publish to a wider audience. This is usually done by publishing in a journal or magazine.

Although getting published is a long and difficult process, it is well worth the effort. Besides enhancing your resume, publication opens your work up to review and criticism by the scientific community. This will provide you with good feedback for future experiments.

As for company management. . .

When managing an experiment, you must take responsibility for assuring that conventional experimental practices are followed. Each experiment should be approved in writing, along with plans on how the results will be applied to the competitive position of the company.

Encourage each experimenter to keep complete records for each experiment. It would be best if you audited those records from time to time. Also, experimenters should confirm their results through a second experiment or with pilot runs.

Taking these steps will prevent many of the problems that plague anyone doing research. As we said in the beginning, experimentation is rigidly structured. This includes the management of experimenting. A department that is less than rigid risks missing data, erroneous results, cheating, and the slowly diminishing faith of the company.

Presentation of experimental results

Every experiment should require the presentation of its results to an audience of some sort. Even if that audience is just your boss, it is important to share your findings with someone who can take action.

There are three media for the presentation of the results: written report, oral presentation, and publication.

All presentations must have the following elements. Each medium of presentation involves presenting these elements in different ways.

1. statement of the problem
2. background
3. method
4. results
5. conclusions

The written report

In a written report, these five elements are presented in the order given above. The written report documents your efforts. Keep the text simple and to the point. One or two illustrations may reinforce your main points. Raw data, detailed analysis, and other lengthy portions should be relegated to the appendices. You are trying to show a business audience quickly and accurately why your experimental results are important.

All experiments should end with a written report. You never know who will need these results or when they will need them. A written report will preserve your information.

The oral presentation

In many cases, you will also be required to make an oral presentation of your results. A successful oral presentation depends on judging the education and needs of your audience. Top management will want facts, and the bottom line in plain English. A general audience of workers and supervisors will want to know how the results will affect their jobs.

In an oral presentation, you present your conclusions first. This will capture the audience's attention quickly and should hold it if the rest of your presentation is less than five minutes long. Thus, a good oral presentation is a quick statement of your conclusions followed by a brief review of how you reached these conclusions.

If any illustrations are necessary, restrict yourself to a single large figure that is shown during the entire presentation. Do not use the ANOVA Table. A better alternative would be an effect diagram. Your preparation for the presentation should involve cutting out everything you feel is not absolutely necessary.

At the end of the presentation, take questions. Make your answers as straightforward as possible.

Publication

Eventually, every experimenter who is serious about his or her work should attempt to publish the results of a particularly interesting or original experiment. Publication exposes you to a peer group of other experimenters. Thus, it forces you to go beyond your best work and to attempt perfection.

Trade journals, scientific periodicals, and other sources of publications are extremely strict with your studies. In most cases, you will have to follow a set of guidelines, such as the guidelines published in the *Journal of Quality Technology* or the American Psychological Association's publication guide.

The order of presenting the required elements will depend on the particular publication you choose. Before you choose to publish, please take the time to read several publications in which you believe your experiment belongs. Editors are very busy people, and you will increase your chances if you research your target publications first.

To get the publication process started, merely write a letter to the editor of the publication. Briefly describe your work and ask for a set of their guidelines. Once you have the guidelines, write a good draft copy of your experiment and submit it to the editor for comments. If the work looks promising, the editor will send you further instructions.

Conclusion

Probably the most important aspect of the presentation to remember is that your appearance and the appearance of your writing will affect how the information is accepted. Therefore, please take pains to write in plain and direct language and to print your reports with the best quality possible. Desktop publishing makes near-typeset quality easily obtainable for anyone.

In your personal appearance, be brief and to the point. Feel free to dress up a little for an oral presentation. The look of neatness and organization adds to the credibility of your work. And, whatever you do, relax and enjoy yourself during a presentation.

Definitions of Common Terms
Used by Experimenters

Assignable cause A source of variation in results that can be associated with a particular factor.

Attribute data Qualitative data that can be counted, such as the number of defects in a product.

Blocking The division of the experimental design into "blocks" or sets of homogeneous conditions (such as units from the same production run, shift, or day) under which the experiment can be conducted.

Common cause A source of variation that affects all of the data under study.

Confounding Combining the effect of one factor with the interaction or effect of another factor.

Control group The experimental subjects that receive no exposure to experimental variables.

Correlation The degree of association between two variables.

Dependent variable The response predicted by an experiment.

Degrees of freedom Usually, the number of independent comparisons available to estimate a specific parameter. Also thought of as the amount of information contained within a sample. Usually one less than the number of numbers.

Experiment A structured investigation. An experimenter selects and controls specific factors to study their response on a quality characteristic.

Experimental combination The settings (or levels) of all factors for one run of the experiment. Also known as treatment combination.

Experimental design The arrangement in which an experiment is to be conducted. It includes the selection of levels for one or more factors or factor combinations.

Experimental error Variations in the response of an experiment caused by extraneous factors. Experimental error adds a degree of uncertainty to the results of an experiment.

Factor An independent variable under experimental control (see also **independent variables**).

Independent variables The variables that are expected to change the results of the experiment.

Interaction When the effect of one factor depends on the settings of another factor. In other words, interaction exists if two factors working alone create little effect, but create a significant effect when placed together.

Level A particular setting of a factor.

Main effect A comparison of the responses at each level of a factor averaged over all levels of other factors in the experiment.

Population The totality of units or members in a group under consideration.

Random sample All members of a population have an equal chance of being selected.

Regression Using one or more independent variables to predict changes in a dependent variable.

Reliability The ability to repeat an experiment and obtain similar results.

Repetitions Repeated samples taken during one set of experimental combinations.

Replication Repeating all of the experimental combinations being tested in an experiment.

Response Outcome variable being studied (see **dependent variable**).

Sample A selected subgroup of a population used to gain information about the population.

Treatment Applying a single combination of the factor levels to the experimental unit.

Treatment combination (See **experimental combination**).

Treatment group The experimental subjects that receive exposure to a variable.

Type I error (Alpha Error) Rejecting a null hypothesis that was actually true.

Type II error (Beta Error) Accepting a null hypothesis that was actually false.

Validity The "correctness" of an experiment's design to determine the effect of various factors.

Variable (See **factor**).

Variable data Quantitative information, where measurement using a variable scale is possible.

Recommended Readings

Barker, Thomas. (1985). *Quality by Experimental Design*. Marcel Dekker, Inc, New York.

Box, George, and Draper, N. R. (1987). *Empirical Model-Building and Response Surfaces*. John Wiley and Sons, New York.

Box, George, and Draper, N. R. (1969). *Evolutionary Operation–A Statistical Method for Process Improvement*. John Wiley and Sons, New York.

Box, George, Hunter, W. G., and Hunter, J. S. (1978). *Statistics for Experimenters*. John Wiley and Sons, New York.

Clements, Richard R. (1988). *Statistical Process Control and Beyond*. Robert E. Krieger, Inc., Melbourne, FL.

Davies, O. L., editor (1947). *Design and Analysis of Industrial Experiments*. Oliver and Boyd, London.

Davies, O. L., editor. (1947). *Statistical Methods in Research and Production*, Oliver and Boyd, London.

Emory, Robert (1980). *Business Research Methods*. Richard D. Irwin, Inc., Homewood, IL.

Ishikawa, K. (1976). *Guide to Quality Control*. Asian Productivity Organization, Tokyo.

Juran, Joseph. (1964). *Managerial Breakthrough*. McGraw-Hill, New York.

Juran, J. M., and Gryna, Frank M., Jr. (1980). *Quality Planning and Analysis*. McGraw-Hill, New York.

Mason, Robert. (1982). *Statistical Techniques in Business and Economics*. Richard D. Irwin, Inc., Homewood, IL.

Taguchi, Genichi. (1986). *Introduction to Quality Engineering.* Asian Productivity Organization, Tokyo.

Taguchi, Genichi. (1981). *On-Line Quality Control.* Japanese Standards Association, Tokyo.

Taguchi, Genichi, and Yuin-Wu. (1979). *Off-Line Quality Control.* Central Japan Quality Control Association, Tokyo.

Journals of Reference

Technometrics

Biometrika

Journal of Quality Technology

Journal of the American Statistical Association

Appendix A: Critical T-values
Percent Confidence

		One-Tail 95%	Two-Tail 97.5%	99.5%
One-Tail		95%	97.5%	99.5%
Two-Tail		90%	95%	99%
df	1	6.314	12.706	63.657
	2	2.920	4.303	9.925
	3	2.353	3.182	5.481
	4	2.132	2.776	4.604
	5	2.015	2.571	4.032
	6	1.943	2.447	3.707
	7	1.895	2.365	3.499
	8	1.860	2.306	3.355
	9	1.833	2.262	3.250
	10	1.812	2.228	3.169
	11	1.796	2.201	3.106
	12	1.782	2.179	3.055
	13	1.771	2.160	3.012
	14	1.761	2.145	2.977
	15	1.753	2.131	2.947
	16	1.746	2.120	2.921
	17	1.740	2.110	2.898
	18	1.734	2.101	2.878
	19	1.729	2.093	2.861
	20	1.725	2.086	2.845
	21	1.721	2.080	2.831
	22	1.717	2.074	2.819
	23	1.714	2.069	2.807
	24	1.711	2.064	2.797
	25	1.708	2.060	2.787
	26	1.706	2.056	2.779
	27	1.703	2.052	2.771
	28	1.701	2.048	2.763
	29	1.699	2.045	2.756
	30	1.697	2.042	2.750
	31	1.695	2.039	2.744
	32	1.694	2.037	2.738
	33	1.692	2.034	2.733
	34	1.691	2.032	2.728
	35	1.690	2.030	2.724
	40	1.684	2.021	2.704
	50	1.676	2.009	2.678
	100	1.660	1.984	2.626
	150	1.655	1.976	2.609

Appendix B: Critical F-values at 95% Confidence

df of numerator

df	1	2	3	4	5	6	7
1	161.40	199.50	215.70	224.60	230.20	234.00	236.80
2	18.51	19.00	19.16	19.25	19.30	19.33	19.35
3	10.13	9.55	9.28	9.12	9.01	8.94	8.89
4	7.71	6.94	6.59	6.39	6.26	6.16	6.09
5	6.61	5.79	5.41	5.19	5.05	4.95	4.88
6	5.99	5.14	4.76	4.53	4.39	4.28	4.21
7	5.59	4.74	4.35	4.12	3.97	3.87	3.79
8	5.32	4.46	4.07	3.84	3.69	3.58	3.50
9	5.12	4.26	3.86	3.63	3.48	3.37	3.29
10	4.96	4.10	3.71	3.48	3.33	3.22	3.13
11	4.84	3.98	3.59	3.36	3.20	3.09	3.01
12	4.75	3.88	3.49	3.26	3.11	3.00	2.91
13	4.67	3.81	3.41	3.18	3.02	2.91	2.83
14	4.60	3.74	3.34	3.11	2.96	2.85	2.76
15	4.54	3.68	3.29	3.06	2.90	2.79	2.71
16	4.49	3.63	3.24	3.01	2.85	2.74	2.66
17	4.45	3.59	3.20	2.96	2.81	2.70	2.61
18	4.41	3.56	3.16	2.93	2.77	2.66	2.58
19	4.38	3.52	3.13	2.89	2.74	2.63	2.54
20	4.35	3.49	3.10	2.87	2.71	2.60	2.51
21	4.32	3.47	3.07	2.84	2.68	2.57	2.49
22	4.30	3.44	3.05	2.82	2.66	2.55	2.46
23	4.28	3.42	3.03	2.80	2.64	2.53	2.44
24	4.26	3.40	3.01	2.78	2.62	2.51	2.42
25	4.24	3.38	2.99	2.76	2.60	2.49	2.40
30	4.17	3.32	2.92	2.69	2.53	2.42	2.33
35	4.12	3.27	2.87	2.64	2.48	2.37	2.28
40	4.08	3.23	2.84	2.61	2.45	2.34	2.25
50	4.03	3.18	2.79	2.56	2.40	2.29	2.20
75	3.97	3.12	2.73	2.49	2.34	2.22	2.13
100	3.94	3.09	2.70	2.46	2.30	2.19	2.10
150	3.90	3.06	2.66	2.43	2.27	2.16	2.07

df of denominator

Appendix B: Critical F-values at 95% Confidence

df of numerator

df	8	9	10	12	15	20	25	30
1	238.90	240.50	241.90	243.90	245.90	248.00	249.30	250.10
2	19.37	19.38	19.40	19.41	19.43	19.45	19.46	19.46
3	8.84	8.81	8.79	8.74	8.70	8.66	8.63	8.62
4	6.04	6.00	5.96	5.91	5.86	5.80	5.77	5.75
5	4.82	4.77	4.74	4.68	4.62	4.56	4.52	4.50
6	4.15	4.10	4.06	4.00	3.94	3.87	3.84	3.81
7	3.73	3.68	3.64	3.58	3.51	3.44	3.40	3.38
8	3.44	3.39	3.35	3.28	3.22	3.15	3.11	3.08
9	3.23	3.18	3.14	3.07	3.01	2.94	2.89	2.86
10	3.07	3.02	2.98	2.91	2.84	2.77	2.73	2.70
11	2.95	2.90	2.85	2.79	2.72	2.65	2.60	2.57
12	2.85	2.80	2.75	2.69	2.62	2.54	2.50	2.47
13	2.77	2.71	2.67	2.60	2.53	2.46	2.41	2.38
14	2.70	2.65	2.60	2.53	2.46	2.39	2.34	2.31
15	2.64	2.59	2.54	2.48	2.40	2.33	2.28	2.25
16	2.59	2.54	2.49	2.42	2.35	2.28	2.23	2.19
17	2.55	2.49	2.45	2.38	2.31	2.23	2.18	2.14
18	2.51	2.46	2.41	2.34	2.27	2.19	2.14	2.11
19	2.48	2.42	2.38	2.31	2.23	2.16	2.11	2.07
20	2.45	2.39	2.35	2.28	2.20	2.12	2.07	2.04
21	2.42	2.37	2.32	2.25	2.18	2.10	2.04	2.01
22	2.40	2.34	2.30	2.23	2.15	2.07	2.02	1.98
23	2.38	2.32	2.28	2.20	2.13	2.05	2.00	1.96
24	2.36	2.30	2.26	2.18	2.11	2.03	1.98	1.94
25	2.34	2.28	2.24	2.16	2.09	2.01	1.96	1.92
30	2.27	2.21	2.16	2.09	2.02	1.93	1.88	1.84
35	2.22	2.16	2.11	2.04	1.96	1.88	1.82	1.79
40	2.18	2.12	2.08	2.00	1.92	1.84	1.78	1.74
50	2.13	2.07	2.03	1.95	1.87	1.78	1.73	1.69
75	2.06	2.01	1.96	1.88	1.80	1.71	1.65	1.61
100	2.03	1.98	1.93	1.85	1.77	1.68	1.62	1.57
150	2.00	1.94	1.89	1.82	1.73	1.64	1.58	1.54

df of denominator

Appendix B: Critical F-values at 95% Confidence

df of numerator

df	50	75	100	150
1	251.80	252.60	253.00	253.50
2	19.48	19.48	19.49	19.49
3	8.58	8.56	8.55	8.54
4	5.70	5.68	5.66	5.65
5	4.44	4.42	4.40	4.39
6	3.75	3.73	3.71	3.70
7	3.32	3.29	3.28	3.26
8	3.02	2.99	2.98	2.96
9	2.80	2.77	2.76	2.74
10	2.64	2.60	2.59	2.57
11	2.51	2.47	2.46	2.44
12	2.40	2.37	2.35	2.33
13	2.31	2.28	2.26	2.24
14	2.24	2.20	2.19	2.17
15	2.18	2.14	2.12	2.10
16	2.12	2.09	2.07	2.05
17	2.08	2.04	2.02	2.00
18	2.04	2.00	1.98	1.96
19	2.00	1.96	1.94	1.92
20	1.97	1.93	1.91	1.89
21	1.94	1.90	1.88	1.86
22	1.91	1.87	1.85	1.83
23	1.88	1.84	1.82	1.80
24	1.86	1.82	1.80	1.78
25	1.84	1.80	1.78	1.76
30	1.76	1.72	1.70	1.67
35	1.70	1.66	1.64	1.61
40	1.66	1.61	1.59	1.56
50	1.60	1.55	1.52	1.50
75	1.52	1.47	1.44	1.41
100	1.48	1.42	1.39	1.36
150	1.44	1.38	1.34	1.31

df of denominator

Appendix C: Factors for Orthogonal Polynomial Equations

Number of Levels	W_1	W_2	W_3	W_4	W_5	λ^2_s	λs
2	$b_1 = -1$	1				2	1
3	$b_1 = -1$	0	1			2	2
	$b_2 = 1$	-2	1			6	2
4	$b_1 = -3$	-1	1	3		20	10
	$b_2 = 1$	-1	-1	1		4	4
	$b_3 = -1$	3	-3	1		20	6
5	$b_1 = -2$	-1	0	1	2	10	10
	$b_2 = 2$	-1	-2	-1	2	14	14
	$b_3 = -1$	2	0	-2	1	10	12
	$b_4 = 1$	-4	6	-4	1	70	24

Appendix D: Documenting an Experiment

One of the most critical steps an experimenter takes is to document an experiment as completely as possible. Every step of the investigation must be accompanied by written records.

Initial proposal

Before an experiment begins, the experimenter must conduct a preliminary investigation of the situation. If the initial data indicate that an experiment is warranted, then this data must be presented in a written proposal to management.

As an example, an engineer may receive a memo telling him to investigate the short life cycle of electrical circuits. His initial investigation may uncover that several components of the circuit are of questionable quality and that the wave solder technique is being blamed for most of the problems.

He would then summarize the quality audit records and the results of his own test in a proposal. Such a proposal would point out that the manufacturing department has identified several possible causes. To determine the best solution to the problem, it will be necessary to uncover the actual causes or combination of causes for short product life. Therefore, an experiment is proposed.

Logbook

The logbook is a day-by-day account of the actions taken in the experimental process. Traditionally, you should have a logbook for each experiment.

A typical entry in the logbook might be,

Dec. 7, 1988

Hal Walker and I interviewed the following people for their opinions concerning the plastic injection problem.

Steve Leeward, plastics engineer. He believed that injection pressure and regrind were major causes of the problem.

Frank Terms, manager of production. He voiced a concern that the experiment must give results that are practical and effective. He requested a trial run of any results as confirmation.

Iva Mercedes, president. She showed us several letters from customers asking if we were conducting experiments. We have been told to submit documentation as proof.

...and so on.

The logbook allows you to review records of your actions when questions arise about the conditions under which you conducted particular tests. You can also record comments made by the people you interview as part of the investigation phase of experimentation.

The experiment

During the experiment, you will use several forms of documentation. The first is the actual schedule of experimental runs. For example, a 2^3-factorial experiment suggests eight combinations, but you must randomize these and convert them into written instructions. Therefore, the operator will see the factorial experiment as a set of instructions like the following.

Plastic Injection Machine Study

It is critical that each of the settings on the machine and the materials used be held constant, except for the following.

Run #1:

Set the injection speed to 3.5 seconds.

Set the mold temperature to 340°.

Set the mold delay to 2 seconds.

Confirm that the temperature has reached the set level. Then mold three sets of parts and mark them "number one."

Run #2:

Set the injection speed to 3.5 seconds.

Set the mold temperature to 340°.

Set the mold delay to 3 seconds.

…and so on.

Each of the parts is marked by its experimental run number. The experimenter should keep a schedule of the actual combination of factors for each run number. In that way the experimenter can find the data that correspond to the experimental run.

Another set of documents to keep is the record of the conditions under which you ran the experiment. Even in a laboratory it is sometimes difficult to control all of the variables. Therefore, you should note in the experimental logbook the details on the settings used, the environmental conditions, and any changes in the experimental design.

A sample entry may be:

September 25

Conducted a test of differences between molding cavities on the machine. Marked parts by which mold they belonged to and measured parts for shrinkage.

Made four parts for each cavity for a total of 48 parts. John Dygert was the operator. Temperature in the area was 78° F with 90 percent relative humidity.

Settings for the test:

Mold temperature 400°

Mold delay 3 seconds

Injection speed 3 seconds

Material used: Stock #234-D

Note how even the smallest details are recorded. At this point the experimenter doesn't know whether humidity in the room is a factor. Therefore, the current humidity is recorded. If humidity turns out to be important, we can always perform a Posttest study of the results.

Communication with participants

An often-overlooked area of experimentation is the social and political impact it has on workers and technicians. Therefore, whenever you plan an experiment, you should issue three types of memos.

The first is a memo to anyone with the slightest interest in the experiment. It should outline what is being studied, why it is important, and how it will be studied. The need for everyone's cooperation should be mentioned gently.

The second memo should go directly to the participants in the experiment. This includes the operators of the machine and the engineers who are participating. This memo should thank these people for their cooperation in the experiment. Also, place copies of this memo in the participants' personnel files. A machine operator who receives recognition for participating in an experiment will be more willing to take part in future studies.

Issue the third memo after obtaining the results of the experiment and taking corrective actions. This memo should again thank everyone who participated. Then it should outline what was learned and how it will benefit the company. After all, good feedback and communication will prevent many of the political problems an experiment might introduce to the manufacturing environment.

Finally

Finally, document the results of an experiment as completely as possible. A final report, the logbooks, and the memos related to the experiment should be gathered together and filed. If, in the future, a similar experiment is planned, a new experimenter can pull this file and get the complete picture on what was learned in the past. This will save repetition of effort or unnecessary searching for clues.

Index

 **Printed on Recycled Paper**